Inland Water Resources
India

Inland Water Resources India

Volume 2

Editors
M.K. DURGA PRASAD
&
P. SANKARA PITCHAIAH
Dept. of Geology
Nagarjuna University
Nagarjunanagar-522 510
Andhra Pradesh

1999
Discovery Publishing House
New Delhi-110 002

First Published-1999

ISBN 81-7141-511-3 (Set)

Published by :
DISCOVERY PUBLISHING HOUSE
4831/24, Ansari Road, Prahlad Street,
Darya Ganj, New Delhi-110 002 (*INDIA*)
Phone: 3279245
Fax: 91-11-3253475

Laser Typeset at:
BHARGAVA PRINTO
1643, Dakhni Rai Street,
Darya Ganj, Delhi-2

Printed at:
Arora Offset Press,
Delhi-110092.

PREFACE

In India, the vast Inland Water Resources are ill managed and improperly utilised. As such, the sustainable clean waters and their resources are becoming rare commodities! It is a fact that the optimal and sustainable utilisation of these waters and their resources require scientific and technical input from various disciplines. Hence, an attempt is made to gear up the valuable scientific information collected by the scientific personnel from all over India belonging to various disciplines such as Agriculture, Aquaculture, Botany, Chemistry, Civil Engineering, Fisheries, Geography, Geology, Horticulture, Physics, Remote Sensing, Toxicology and Limnology. We hope the cluster of knowledge presented in these volumes will be useful to the reader.

M.K. Durga Prasad
P. Sankara Pitchaiah

Contents

Section III
Methodology

Section IV
Pollution Studies

Section V

Water Development and Management

Analysis of Groundwater Flow Regime through Simulation Models

M. THANGARAJAN, V.S. SINGH
AND
K. SUBRAMANIEM

Introduction

Groundwater is one of the most important and widely distributed natural resource which is considered as supplemental resource to surface water to meet the domestic, agricultural and industrial requirements. Unplanned development of groundwater often results in adverse effect of continual decline of potentiometric surface, water quality degradation and excessive consolidation of different subsurface state manifesting as land subsidence. Obviously, it is not possible to carry out experiments and tests in the aquifer itself (with prohibitive cost) in the order to determinc whether the available resource can meet the demand in future without any adverse effect. Usually, simulation models are used to interpret pumping test data in large diameter wells and assess the response of the aquifer to different operating scenarios in future.

Occurrence of Groundwater in Hard Rock Areas

The 'in situ' resource in the groundwater reservoir in the hard rock terrains is dominated by the vadose zone through which the water level fluctuates. However, the fracture porosity of igneous and metamorphic rocks is small and the existence of fracture zone alone is not sufficient for large productivity of wells, conductive fractures must have connection with some region of substantial groundwater storage (a thick mantle of weathered rock; a surface water body).

Hard rock aquifers assume importance in view of their large aerial extent and compiled nature. About two-third of India's geographical area is underlain by hard rocks. The crystallines and meta-sediments form the hard rock aquifers. These aquifers are highly complex and non-homogeneous both laterally and vertically.

Recharge and discharge areas are separated by regions of horizontal flow. Groundwater flow systems may be shallow or deep. In shallow systems, which are characteristic of hard rock terrains, the flow may penetrate to few tens of metres below land surface, and the length of flow path may be from few tens to few hundreds of metres. The shallow systems are sensitive to seasonal changes in precipitation and evapotranspiration.

In view of the vast aerial distribution of the hard rocks and as groundwater occurs only in the secondary intestices, pores and void spaces in them, the existing concepts of soft rock hydrology are barely adequate and development of concepts specific to hard rocks becomes imperative.

Parameter Estimation through Analysis of Pumping Test Data from Large Diameter Wells in Crystalline Rocks

In crystalline rocks the upper weathered mantle forms shallow aquifer which generally is tapped by large diameter wells for the extraction of domestic as well as irrigation purposes. It is necessary to estimate the aquifer parameters of the shallow aquifer zone in order to determine the total groundwater potential in an area. Although the pump fitted in large diameter well offer cost-effective means to estimate the aquifer parameters, the interpretation of the test data is brought with several ambiguities. It is often found that the aquifer response is negligible and the abstraction rate varies during the pumping phase. These restrict the use of conventional type curves for the estimation of aquifer parameters. Under such circumstances, the pumping phase and recovery phase has been used to estimate the aquifer parameters utilizing the numerical method.

Methodology

The method consists of discrediting entire pumping duration into a number of equal time steps (Δt). The aquifer response is calculated for each time step through convolution of the abstraction with the impulse response function. The abstraction rate during each time step is assumed to be constant, while it could take different values during various time steps.

The abstraction Q from a large diameter well during each time step comprises abstraction Q^w from the well storage and the contribution Q^a from the aquifer. Thus, the abstraction rate after 'n' time steps is given as :

$$\sum_{i=1}^{n} Q = \sum_{i=1}^{n} Q^w + \sum_{i=1}^{n} Q^a \quad (1)$$

The net drawndown in the well (s) may be expressed as :

$$S = \frac{\Delta t}{\pi r_w^2} \sum_{i=1}^{n} Q_i^w \tag{2}$$

where r_w is the radius of the well.

Similarly the drawdown at the well face in the aquifer may be expressed as :

$$S_a = \frac{1}{4\pi T} \sum_{i=1}^{n} Qi\, W(n\text{-}i+1) \tag{3}$$

where T is transmissivity and W the impulse response of the aquifer.

In case of no seepage face developed during the test the drawdowns represented by equation (2) and (3) will be equal. Therefore, one can easily find out the unknowns Q^a and Q^w from the equations (1), (2) and (3) and finally the drawdown in the well is calculated using either equation (2) or (3).

These calculated drawdowns are then compared with the observed drawdowns in the well. The aquifer parameters are progressively modified in order to get a close match between the observed and calculated time-drawdown/recovery. The best fit time-drawdown/recovery curve gives the representative aquifer parameters.

Case Study : Palar River Basin

The part of the Palar river basin (Long. 79° 15′–80°15′ and Lat. 12°25′–15°00′) which falls in South Arcot District of Tamil Nadu, is occupied by crystalline rocks. (Fig. 24.1). In spite of large catchment, the river Palar is dry during most part of the year. There are several irrigation tanks on the sides of Palar river. Many townships draw their water supply from subsurface flow of river bed.

The Archaean crystalline rocks consist of gnesises, granites and charnockites. These are at places overlain by alluvium. At places these are interested by dolerite dykes and quartz veins. Since these dykes are of highly resistive nature, they form the central ridge of long crested hills.

The crystalline rock generally is covered by weathered mantle of coarse sandy and clayey material of about 15 m thickness. Underneath the weathered mantle lies fractured and jointed crystalline rock. Groundwater occurs in weathered as well as in the fractured zones. The large diameter wells are most effective means to exploit groundwater in the poorly

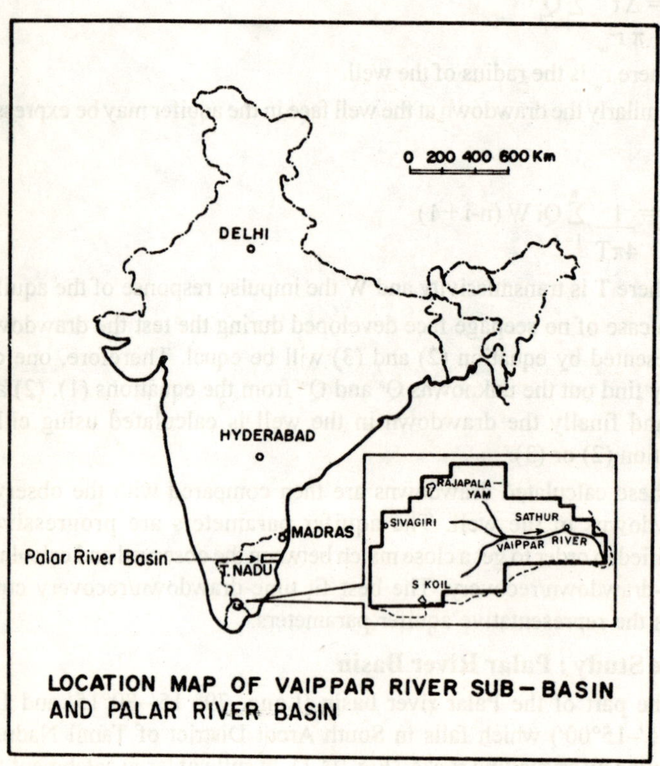
0 200 400 600 Km
DELHI
HYDERABAD
MADRAS
Palar River Basin
T. NADU
RAJAPALA-YAM
SIVAGIRI
SATHUR
VAIPPAR RIVER
S KOIL
LOCATION MAP OF VAIPPAR RIVER SUB-BASIN AND PALAR RIVER BASIN

Fig. 24·1

permeable aquifer of crystalline rock. The depth of these wells vary from shallow 3-5 m to 16-20 m. At few places the water level is at shallow depth, 2-3 m. below ground level (bgl) whereas the other places it is as deep as 10-15 m. bgl.

Pumping Test

As most of the dug wells are fitted with pumps, they offer cost-effective means to conduct pumping test for the estimation of hydrogeological parameters. Some of the pumping tests from the area are selected for the interpretation. The details of these tests are given in Table 24.1. The diameter of the dug wells vary from 3.5 to 10.4 m. whereas the depth from 3.65 to 15.3 m. bgl.

Table 24.1 : Details of Pump tests

Well No.	*Diameter (m)*	*Depth (m)*	*pumping Rate (cu. m/d)*	*pumping Period (min)*	*Maximum drowndown (m)*	*Recovery Period (min)*
1.	6.00	3.65	935	90	2.12	480
2.	8.10	15.90	1962*	120	3.1	120
3.	5.38	10.70	815	70	1.75	105
4.	9.50	10.50	583	150	0.62	300
5.	5.34	9.50	257*	150	1.25	150
6.	6.20	13.40	382	120	1.22	240
7.	3.50	15.23	857*	90	1.03	210
8.	5.00	15.30	1102*	190	3.45	190
9.	5.16	8.03	700	75	1.52	150
10.	10.40	5.25	1222*	100	0.81	320
11.	5.40	8.90	692	55	1.07	110

* variable abstraction rate

The total duration of the test varies from 70 minutes (min.) to 190 min. The maximum drawdown observed in these wells are given in Table 24.1. In some of the tests the discharge rate was not constant. Also, due to poor permeability of the formation, the pumping phase data does not show any appreciable aquifer response. Thus, the conventional type curves cannot be used to interpret the test data. The numerical method (Singh and Gupta, 1986) which takes into account such features, has been adopted to interpret the test data.

Interpretation and Results

The pumping phase as well as the recovery phase data have been considered for the estimation of transmissivity (T) and storage coefficient (S) of the formation. In the test, where the abstraction rate was found variable (see Table 24.1), the actual abstraction rate at constant interval of time from the observed data was taken into account.

The initial guess values of T and S were considered (depending upon the hydrogeological conditions) to calculate the time-drawdown/recovery. These are compared with observed time-drawdown/recovery values. The T and S values are then progressively varied in order to get a close match between observed and calculated drawdown/recovery.

The aquifer parameters as estimated from the pumping test are given in Table 24.2. The transmissivity values were found to vary from 10 to 250 sq m/d whereas the storage coefficient as 0.0001 to 0.07.

Table 24.2 : Aquifer Parameters

Well No.	*Initial guess values*		*Final Estimated Values*	
	T (*sq. m/d*)	*S*	*T* (*sq.m/d*)	*S*
1.	5	0.001	10	0.001
2.	100	0.001	55	0.013
3.	5	0.001	35	0.02
4.	50	0.005	100	0.01
5.	50	0.01	100	0.05
6.	50	0.001	90	0.025
7.	100	0.0001	250	0.001
8.	30	0.002	50	0.01
9.	70	0.005	100	0.01
10.	10	0.005	20	0.07
11.	5	0.005	15	0.001

4. Numerical Simulation of Flow Regime in Vaippar River Basin

Introduction

Numerical simulation of Vaippar river basin in Tamilnadu was taken up with the available hydrological data, in order to understand the dynamics of groundwater flow system and to identify the data gaps for refinement of the model (Thangarajan and Trippler 1984).

Physiography

The study area is located in Ramnad and Tiruneyveli districts of Tamilnadu (Fig. 24.1). It covers about 1350 sq km between the longitudes 77°25'E to 78°E, and latitudes 9°0'N to 9°30'N. The area is drained from west to east by the river Vaippar and a number of tributaries. The general surface slope is 0.002 (Fig. 24.2). The area is semi-arid with seasonal rains during October until January. The mean annual rainfall is about 650 mm. Periodic droughts hit the area due to monsoon failure.

Geology

The area of investigation is a part of the Archaean shield of South India dipping towards the east. Due to the tectonic activities the granite below the phreatic aquifer is fractured with so far unknown main direction and magnitudes of the fractures. The phreatic aquifer of about 20 m thickness consists of highly weathered gneisses and charnockites. The top-soil consists of clay/sand and clay/kankar in the eastern part and sandy loam soil in western part, which is good for paddy, chilli, and cotton crops.

Hydrogeology

The groundwater occurs mostly in the top 20 m weathered zone of gneisses and charnockites. Many bore wells were drilled by Public Works Department of Tamil Nadu in the phreatic aquifer. Almost none of these wells enter below weathered rocks, so that nothing can be said about the size and main direction of the fractures. Due to the topography and the distribution of the surface water system and depth to groundwater varies from 0 to 10 m.

The annual water level fluctuation is normally between 4 to 5 m. Only during drought years that water level may go down by even more than 10 m followed by a very fast recovery of the groundwater table after the first heavy rainfall.

In general, the groundwater discharges towards the rivers and as such to the east. The groundwater velocity is low due to the low transmissivities. As such, only locally, the groundwater head can be influenced by man made activities.

The transmissivity pattern was evaluated from results of 28 short term pump-tests on dug and bore wells provided by Groundwater Wing of Public Works Department, Tamil Nadu. Very low and high transmissivities were avoided. In general, the transmissivities are higher in the west and become low in the discharge region in the east. The storage coefficient

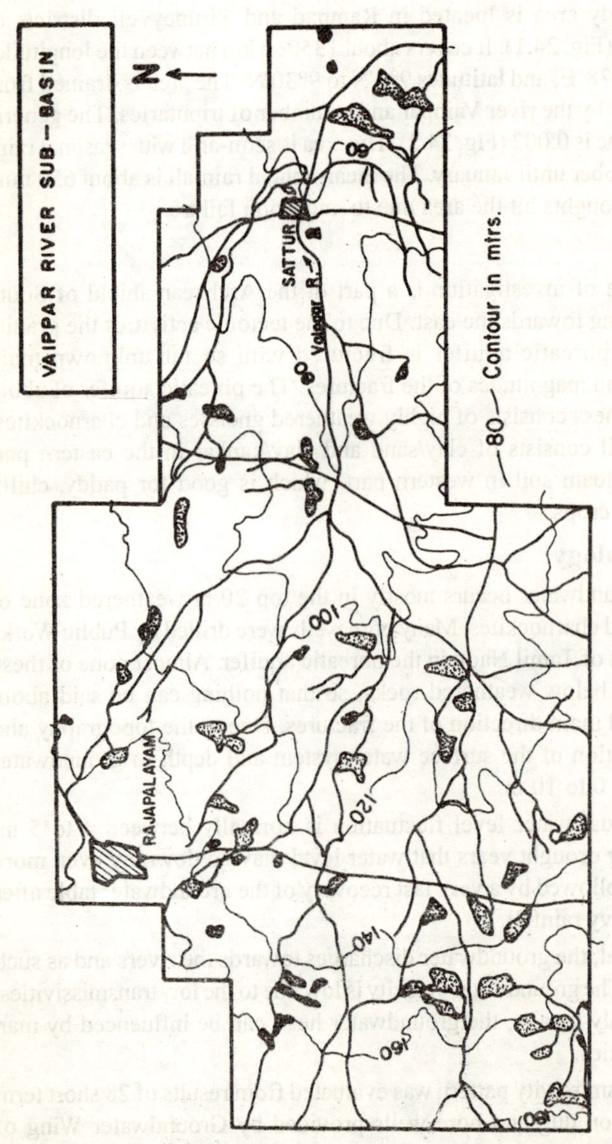

Fig.24.2 Topographic contours given in 20 m intervals

was estimated through pump-tests to be 0.15. It was inferred from the infiltration studies available at few selected places that about 15% of the rainfall recharges to the groundwater and 5% as an additional indirect recharge from surface water system.

THE SIMULATION MODEL

The Concept and Realization of the Model

It is assumed for this study that the flow of water through the weathered material is of Darcian nature. Of course, the understatum hard rocks are fractured and as such groundwater flow along those fractured too. But, since no detail information about the location, size distribution and direction of the fractures are known, they are kept out of this model study. As such, a two dimensional single layer horizontal model for the weathered aquifer only was attempted.

Model Calibration

The model was calibrated for steady state and transient conditions. The initial condition for the model calibration was the water level of January 1973.

Model test runs with constant head boundaries

The starting conditions for the test runs with the steady state model are fixed heads at the boundaries and the rivers. A net recharge of 100 cu.m/day per node point was given as an input to the system. The system was then balanced by groundwater discharge to the rivers. The model result showed that the inner-nodal heads in northern part was 20 m. above the field values. It is for this reason the boundary head at north-west was then fixed at 140 m. This requires an explanation that there should be a fault passing through north-south and affecting the transmissivities locally. This had to be checked in the field and should be incorporated in the refined model. The groundwater contours clearly exhibits that the rivers are effluent and micro-basin characteristics occurs.

Model test runs with lateral inflows/outflows

In the next series of steady state test runs the lateral flows were given all along the boundaries of the model. The total lateral inflow was calculated using hydraulic gradient and transmissivity, as 15,575 cu.m.. Out of this 300 cu.m. discharges laterally at the eastern boundary. Maximum groundwater discharges towards the rivers. During the steady state calibration of the simulation model the initial transmissivity was modified from 25 sq. m. to 50 sq.m. in the south-eastern part of the model

(Fig. 24.3). The comparison of computed and observed water level contours for steady state model is shown in Fig. 24.4.

Transient model calibration

Storativity of 0.15 based on dug-well pumping tests was assumed for the initial transient model calibration. The mean groundwater fluctuation in this area is around 4 to 5 m. Through model calibration the withdrawal rate at each node point was estimated as 9,000 cu.m..

The recharge rate is estimated to be 20% (3/4 direct and 1/4 indirect recharge) of the mean annual rainfall during a three month period.

With these initial assumptions the model could not be in balance, since the withdrawal quantity is 4.8 times larger than the groundwater recharge. Only if the storativity is reduced to 2.5% then the system becomes balanced.

The model was then run for the time of 1973 to 1978 using 20% of the mean annual rainfall as recharge, except for the dry years only 4% of the mean annual rainfall was used.

Both measured and simulated heads are shown in standardised form in Fig. 24.5 [H stan = (h-hm)/sh, where H stan = standardized head (unitless), h = measured or simulated groundwater head, hm = mean of all values on h at a location, sh = standard deviation of h]. As such it is assumed that the parameters of T and S and the recharge and withdrawal quantities used in this simulation model may be not too far away from the reality.

Uncertainties of the Model

In any work with a simulation model for groundwater systems a lack of data is found in either having proper aquifer parameters or input and output quantities or in both. This model study also does not deviate from the above mentioned facts.

The following assumptions were made to realize the model system. These assumptions may mirror the prototype but a field check of the data may ensure the present model concept.

(*i*) Throughout the simulation model the stream beds were fixed at constant heads based on measures from the topographic map. This assumption may not be correct for all the seasons because during drought years the groundwater level may go down below the stream beds. By fixing the river heads in the model, induced infiltration from the streambeds to the aquifer is present during droughts. This may not represent the realistic situation.

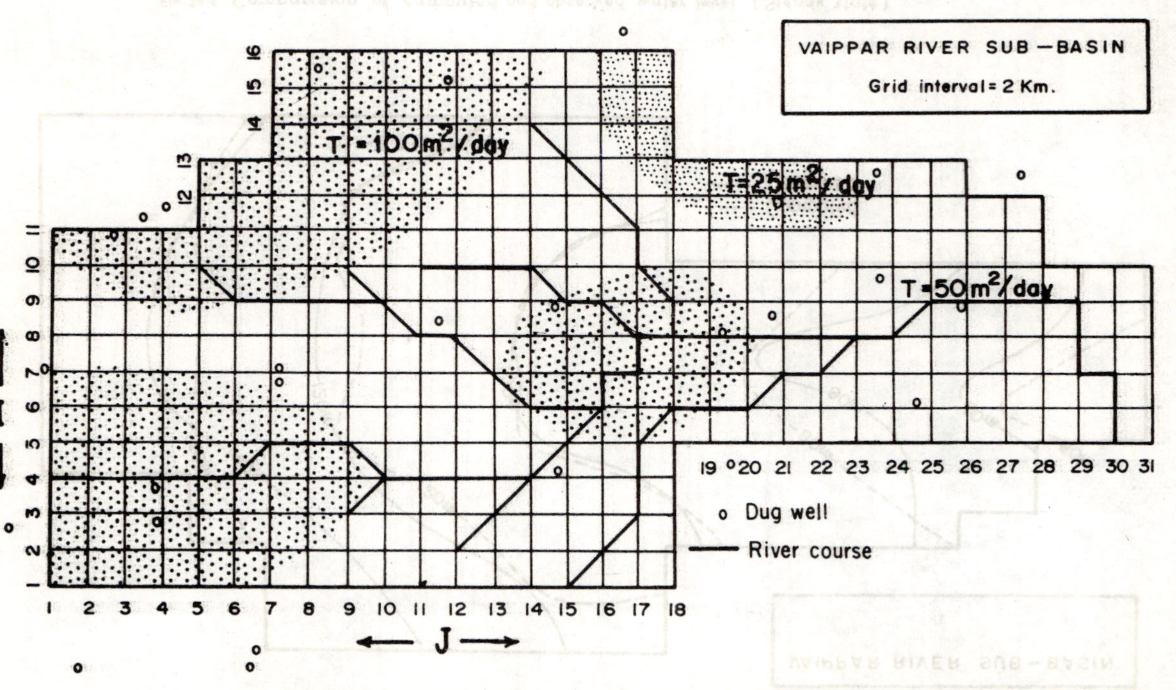

Fig. 24.3 Transmissivity pattern together with model grid

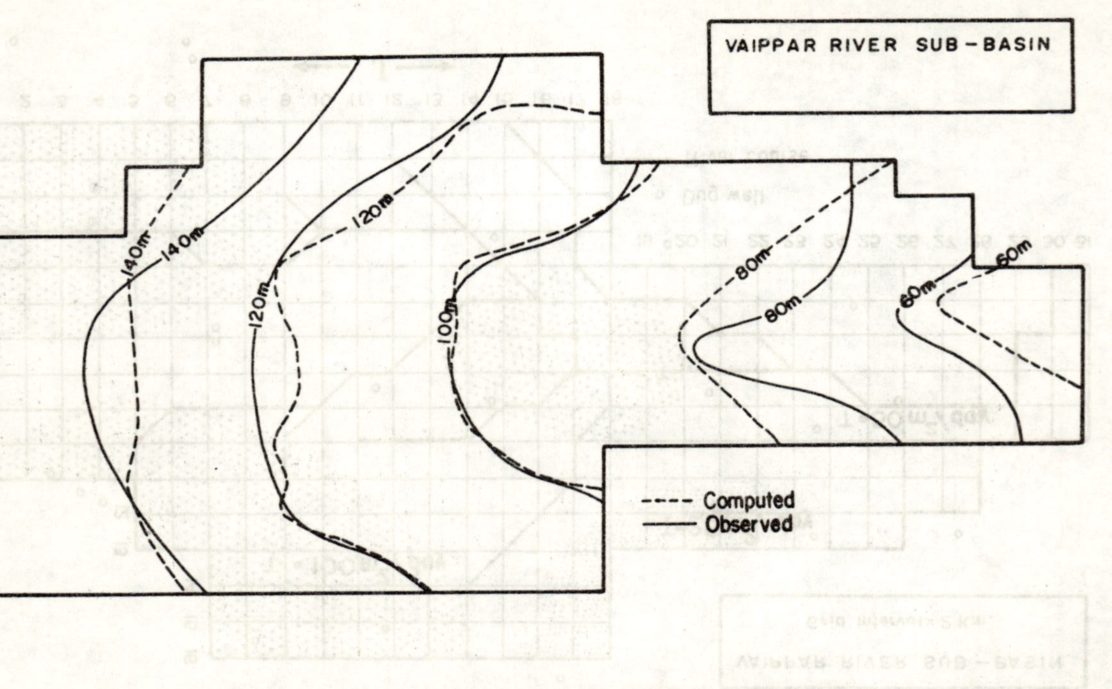

Fig. 24.4 **Comparision of computed and observed water level (Steady state)**

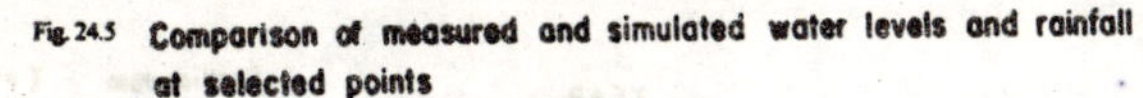
Fig. 24.5 Comparison of measured and simulated water levels and rainfall at selected points

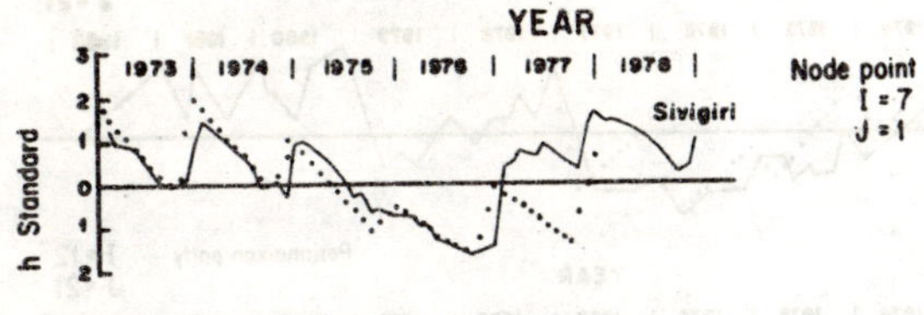

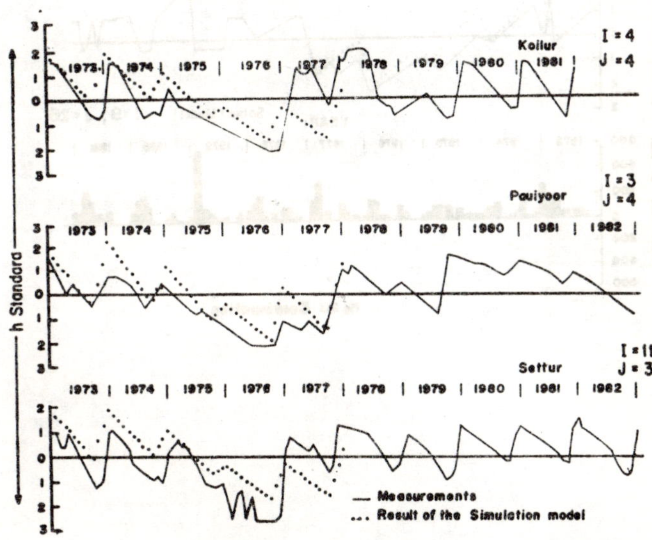

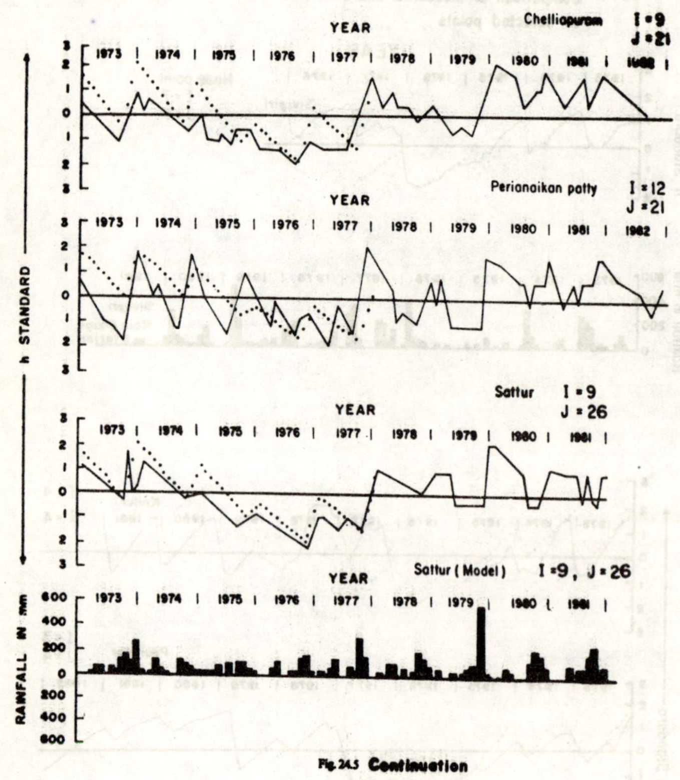

Fig. 24.5 **Continuation**

(*ii*) The total withdrawal rate is not known. It is only estimated that about 1970 cu. m. is pumped out of an area of one node point (4 sq.km.) which comes to about 170 mcm. per year for the total area. This figure of total withdrawal was used in the time depending runs of the simulation model.

(*iii*) The storativity determined through short term dug-well pump test show a wide range from 15% to 2.5%. During the calibration of the model it was found that S = 0.025 is the most appropriate figure for the whole area. The figure of S has to be checked again in the field by means of long term pump tests.

(*iv*) As outlined earlier the interaction of the surface and the groundwater system is important. It is assumed that additional indirect recharge is present during the post monsoon time. No estimate of this figure was introduced to the present model. But, in further model studies it will be necessary to know a real extent and distribution of the tanks and the impounding time.

(*v*) The river flow data and the recharge/discharge of groundwater from/to the rivers are not available at present. These data have to be collected and introduced in further model studies.

(*vi*) The recharge of 15% of the mean rainfall is inferred from infiltration studies in the paddy fields and seepage studies at three irrigation tanks. This figure again has to be verified by injected tracer studies.

Model Results

The results obtained through test runs of the simulation model of the Vaippar river basin are discussed, here.

Results of the Steady State Simulation Model

The concept of a two dimensional horizontal simulation model was verified during the calibration runs with the model system.

The final transmissivity distribution was changed only in the south eastern part of the model to 50 sq. m.. The assumed interaction between the surface water and the groundwater system was introduced into the model, by keeping the groundwater heads constant at the node points of the river. In future model work, time dependent recharge and discharge conditions at node points contact with the surface water system should be introduced instead of constant head conditions.

Result of the Time Depending Simulation Model

In general the computed water level hydrography satisfactorily match with field hydrographs. Deviations of the simulated heads from the measured heads can be explained by local deviations of rainfall from the mean or/and the presence of variable indirect recharge/discharge for groundwater from the surface water system.

Recommendations

The following recommendations have merged out as a result of first simulation model studies. Additional data collection has to be carried out for future refinement of the model.

(*i*) The withdrawal distribution in space and time is unknown. This data is the most essential and may be estimated by an interpretation of the landuse, aerial photographs, satellite-imageries, and detailed water balance study in a micro basin within the area or by the energy consumption used for the pumps.

(*ii*) The surface water system is in close contact with the groundwater system. If the dynamics of the groundwater flow system should be understood then more details have to be known about the surface water system *viz.*

(*a*) the groundwater discharge at selected points in the river should be known,

(*b*) the areal extent of the tanks and ponding time of water at selected tanks should be measured, and

(*c*) continuous groundwater level measurement should be done at selected places (at boreholes only).

(*iii*) Rapid recharge is observed all over the area after long dry periods. The mechanism has to be studied and separated from the gase normal recharge conditions by suing tracer techniques.

(*iv*) The hydrogeological boundary of the area of investigations should be better defined for the refined model.

(*v*) The pattern of fractures has to be found, out of satellite-imageries and aerial photographs.

(*vi*) Locally, aquifer parameters have to be determined. Also the interaction of the phreatic system with the underlain hard rock aquifer has to be determined. The local interaction of the groundwater system with the rivers should be investigated in detail. As such, the following activities are recommended :

(*a*) Long duration pump tests should be carried out both in dug wells (unused for the entire period of the test) and bore-holes tapping the deeper system. If possible, separate observations of the drawdown should be carried out in the deeper and in the phreatic part of the aquifer.

(*b*) Vertical permeabilities and thickness of river beds should be known. The groundwater level fluctuations within the dry stream beds during the dry seasons of the year should be monitored, and

(*c*) The model result requests a zone of low transmissivities between Rajapalayam and Mallipudur region of the study area which may be caused by a fault passing from north to south.

Conclusion

In the hard rock aquifers (Palar basin), the estimation of hydrological parameters becomes difficult from the data obtained through pumping phase from large diameter wells. In the paper, the usage of simulation models to pumping test data, using both pumping and recovery phases has been successfully utilised for better estimation of aquifer parameters.

The present simulation model study of Vaippar river basin mainly intends to realize the aquifer system through calibration by using the available hydrogeological data. It becomes questionable to calibrate a model in the absence of any measurements of the withdrawal rates. In that case an attempt has been made for the model calibration by using subjective interference which may not be far away from the realistic situation. The model has to be recalibrated after collecting the necessary data in the subsequent years.

This study helped to understand and the aquifer system and its dynamics. It is here suggested that a refined database and model concept may be used to study the aquifer response for the optimal usage of the groundwater. For this it is necessary that field work and model work should interface each other.

Acknowledgements

The authors would like to thank the Director, N.G.R.I. for permitting to publish this paper. Authors also thank Dr. C.P. Gupta, Scientist, N.G.R.I. and Dr. K. Trippler, Federal Institute of Geosciences, Hannover, Germany for their keen interest in the present study. Authors are also thankful to the officers of State Groundwater Wing of Public Works Department of Tamil Nadu who provided the necessary field data.

References

Gupta, C.P. and Thangarajan, M. (1990). Management of Groundwater Resources India Using Simulation Models. Water Resources Journal. March-1990, 34-42.

Papodopulos, I.S. and Cooper Jr. H.H. (1967). Drawdown in a well of large diameter. Water Resources Research. Vol. 3, No. 1, 241-44.

Singh, V.S. and Gupta, C.P. (1986). Hydrological parameters estimation from pump tests on a large diameter well. Journal of Hydrology. Vol. 87, 223-32.

Thangarajan, M. and Trippler, K. (1984). Simulation model studies of the phreatic aquifer, Vaippar-River-Basin, South India. B.G.R. Technical Report No. 10,174/84.

25

A Systems Approach to Groundwater Resources Development and Management

A.K. Rastogi

Introduction

The demand of water is continuously increasing in almost all the countries of the world today. This demand growth is attrributed to increasing agriculture areas, emphasis on intensive cropping intensity, rising population in some countries of the world, rapid industrialisation and urbanisation. These demands are met by two basic sources of water supply: Surface water and Groundwater. Surface water sources underwent significant development in the past several decades as these were easier to quantify and quite visible. But all surface water supply schemes depend almost directly on annual rainfall and melting of ice and snow cover which no control can be exercised in the near foreseeable future. As a result for the last few decades there is a growing appreciation for effective planning and management of groundwater systems through its complete aquifer stress response analysis and a planning strategy of its gradual capacity expansion.

A groundwater system involves inputs in the form of surface and artificial recharge, irrigation return flows, seepage from streams, lakes, ponds, canals and sub surface inflows. The output from the system are sub surface outflow, discharges to springs and evapotranspiration. Pumping, injection and artificial recharge schedules of groundwater and surface water are termed as decision or control variables, whereas hydraulic head, pressure head or mass concentration characterise the state variables of a groundwater system. Further the system's parameters are characterised by storage and tranmission properties, leakage factor and dispersion coefficients of the system, and finally, all these terms are interrelated to each other through the governing flow equation of the system.

Many problems normally arise due to indiscriminate use of groundwater over long periods. Thus problems of water logging, sea water intrusion in coastal aquifers, land subsidence, critical lowering of water table and water quality deterioration are often attributed to an inadequate analysis of the response of various aquifer stresses on the groundwater behaviour. The techniques of analytical solutions, flow net analysis, physical models and analog models have largely been used to find solutions of these problems. However, these techniques have limitations in dealing with the real world problems. As a result most of the groundwater problems solved using these techniques are over simplified in terms of aquifer parameters and boundary conditions.

More recently easier access to adequate computing resources have focussed more attention towards digital models which have been used for quantified decision making and for understanding the groundwater flow behaviour in aquifers. Digital models can deal with three classes of groundwater management problems, *viz.*, the prediction (forecasting), parameter estimation (inverse) and instrument (detection) problem which are normally considered in the analysis of groundwater systems (Willis and Yeh, 1987). The tremendous value of digital techniques of solution lies in that they allow to conduct 'thought experiment' along with generation of alternatives.

Following sections briefly review the digital techniques along with groundwater management models for a systematic analysis of groundwater resources.

Digital Models

Digital models are also referred as mathematical model, numerical model, computer model and discretised model by various workers. These models have considerably advanced the idea of a systematic planning and management of groundwater reservoirs. Since groundwater management is mainly concerned with the valuation of hydrologic and economic impact and trade offs associated with development and allocation of groundwater supply to competing water uses and demands, digital models of the groundwater system address these problems in detail and find acceptable solutions to real world of heterogeneous and extensive aquifers with irregular boundaries, various boundary conditions and multiple sources of recharge and discharge.

Digital models of groundwater supply system are based on the response equation of the aquifer system. These equations relate the state variable, *viz.*, groundwater head distribution of the system, to management alternative, such as location, magnitude and duration of groundwater pumping or artificial recharge (Willis and Yeh, 1987). Digital modelling involves discretisation of the governing equation describing groundwater flow in a particular aquifer domain. The resulting algebraic equations are then solved by numerical methods along with the boundary conditions. Numerical models are used either for simulation of the response of an aquifer to a deterministic pattern of groundwater withdrawal and recharge, or for estimating aquifer parameters by employing the historical data of aquifer excitation and response. The former is called a direct or prediction problem and the latter an inverse problem. Willis and Yeh (1987) added one more use of these models for the identification of the recharge or leakage in semi confined or confined aquifers, from the response properties of the aquifer system.They termed this as instrument or detection problem. Various techniques have been used for digital simulation of the aquifer systems. Finite Differences and Finite Elements are the two most powerful and widely used techniques of solving the groundwater flow problems. The following section briefly examines their application in the groundwater systems modelling.

Finite Differences

These are approximate methods of solution in the sense that derivatives at a point are approximated by difference quotients over a small interval. Smith (1978) commented that finite difference methods generally give solutions that are either as accurate as the data warrant or as accurate as is necessary for the technical purposes for which the solutions are required. Finite difference models of the groundwater flow and mass transport equation are based on a Taylor series representation of the time and spatial derivatives. For example, governing equation of a regional groundwater system is given as

$$\partial/\partial x\,(T_x\, \partial h/\partial x) + \partial/\partial y\,(T_y\, \partial h/\partial y) + K_z\,(H_z - h)/m \pm \Sigma Q\delta\,(x - x_w)(y - y_w) = S\, \partial h/\partial t \quad (1)$$

where T_x and T_y are the transmissivity values in the prinicipal axes direction, $\partial h/\partial x$ and $\partial h/\partial y$ are the hydraulic gradients in the x and y direction, K_z and m are hydraulic conductivity and thickness respectively of the semi confined aquitard, h is groundwater head averaged over the vertical, H_z is external head in the overlying aquifer, -Q is the discharge (+ Q recharge) from wth pumping well located at (x_w, y_w), S is the storage coefficient of the aquifer, $\partial h/\partial t$ is the time derivative of the groundwater head and $d(x-x_w, y-y_w)$ is the Dirac delta function

where $\delta(x-x_w), (y-y_w) = 1$ if $x = x_w, y = y_w$

$= 0$ if $x/x_w, y/y_w, x \neq x_w, y \neq y_w$

Using finite differences discretisation principles equation (1) transforms to

$$\left\{\frac{2}{x(I+1)-x(I-1)}\right\}\left[T_x(I,J)\left\{\frac{H(I+1,J)-H(I,J)}{X(I+1)-X(F)}\right\}\right.$$

$$\left.-T_x(I-1,J)\left\{\frac{H(I-J)-H(I\text{-}1,J)}{x(I)-x(I\text{-}1)}\right\}\right]+\frac{2}{\{Y(2H)-Y(J-1)\}}$$

$$\left[T_y(I,J)\left\{\frac{H(I,J+1)-H(I-J)}{Y(J+1)-Y(J)}\right.-T_Y(I,J-1)\left\{\frac{H(I,J)-H(I,J\text{-}1)}{Y(J)-Y(J-1)}\right\}\right]$$

$$+\frac{Ka(I,J)}{M(I,J)}\left\{Ha(I,J)-H(I,J)\right\}+\frac{4Q(I,J)}{\{(x(I+1)-x(I\text{-}1)(y(y+1)-y(y-1)\}}$$

$$=S(I,J)h^{\circ} \qquad (2)$$

where X, Y, T_x, T_y and H(I, J) are spatial coordinates, transmissivity and head values. Other values are explained above and now referred for a general node (I, J) h is the time derivative of the groundwater head which can be written as

$$h^{\circ}=\{H(I,J,T+1)-H(I,J,T)\}/\Delta T \qquad (3)$$

where ΔT is the time step and the two head values refer to specific time period.

River flux, canal seepage, artificial recharge and any other source of input or output can be accounted similar to aquitard leakage term in the discretised equation. Aquifer flow domain is divided into large number of aquifer sub domains each represented by a central node. Therefore algebraic equations like Eq. (2) are obtained for all the nodes within the groundwater system.

Various numerical methods have been used by a large number of workers to obtain the solution of these equations. More important of these Modified Interactive Alternating Direction Implicit (MIADI), Successive Over Relaxation (SOR) and Strongly Implicit Procedure (SIP) schemes have largely been used to solve various problems of groundwater systems behaviours for defined acquifer stresses. Remson *et al*. (1971). Prickett and Lonnquist (1975), Trecoot *et al*. (1976), Rushton and Redshaw (1979), Rastogi (1983), Huyakorn and Pinder (1983) and Willis and Yeh (1987) have considered the above methods in greater details and solved a range of groundwater problems using these numerical methods. A few relevant points of these methods are examined in the following section.

In SOR, also referred as Line Successive Over Relaxation (LSOR) method the solution head values improves one row (or column) at a time. Whether the solution is oriented along rows or column is generally immaterial for isotropic problems but has significant effect on the convergence rate in anisotropic problems. Selection of an optimum value of over relaxation parameter w is important for minimising the computational efforts. A value of 1.6 to 1.9 is commonly recommended, however, more efficient methods to get an optimal value of w are suggested by Remson *et al.* Cooley (1974) for transient problems.

IADI given by Peaceman and Rachford (1955) involve solution of two sets of matrix equations in each iteration using Gaussian elemination technique. Stone (1968) modified IADI to reduce the computation time and storage requirements of Gaussian elimination method. The third technique – Strongly Implicit Procedure is more complex but converges more rapidly compared to LSOR or MIADI methods. Stone (1968) and Weinstein *et al.* (1969) concluded that SIP is a more powerful iterative technique than IADI for most problems. However no definite conclusion regarding the superiority of aforesaid numerical methods can be made. Since it normally requires adequate experience to get well conversed with a particular technique is not always convenient. Since easy access to personal computers with large storage facilities does not pose any serious limitation on computer run time resources, any well practised method is useful for adequate solutions of a real world groundwater problem.

Finite Element Method

The finite element method is another numerical technique for the solution of groundwater flow and quality problems. Heterogenetics and irregular boundaries are handled naturally by finite element method. The method is particularly suited to transient seepage problem in unconfined aquifers. Moreover in FEM the size of the element can be easily varied and rapidly changing state variable of parameter values can be adequately handled. The piecewise continuous representation of the dependant variables and possibly, the parameters of the groundwater system can also increase the accuracy of numerical approximations (Willis and Yeh, 1987).

Galerkin procedure is the most widely used FEM method in literature compared to other weighted residual methods for solving a wide ranging groundwater flow problems. To develop a Galerkin finite element model for two dimensional groundwater in any type of aquifer involving various source and sink terms, the orthogonality condition requires that

$$\Sigma N_k L\{h\} dD_e = 0 \, k \tag{4}$$

where L {h} is the appropriate governing equation of groundwater flow in the flow domain which is divided by various elements (e) and summation is over all elements of the system, N_k are the basis functions and D_e is the elemental domain. Selection of proper basis function for the elements (triangular or quadrilateral) within the aquifer domain is important for the accuracy and stability of the Galerkin finite element method.

Each term in eq (4) generates an element matrix which depends on the basic function and system's parameter. Summation of all such element matrix leads to a global matrix which is a set of algebraic equations as obtained in finite difference method. The global matrices typically have a larger number of zero coefficients (sparse matrices). All the non zero coefficients are confined, however, within a narrow band in the global equations. Therefore, the hydraulic or water quality response equations of the groundwater system are obtained by numerically transforming the systems equations using finite elements. These equations are systems of simultaneous equations and can be expressed as

$$A d\text{Ø}/dt + B\text{Ø} + g = 0 \tag{5}$$

$$\text{Ø}(o) = \text{Ø}_o \tag{6}$$

where the A and B coefficient matrices are functions of the system's hydraulic or quality parameter and the basis functions. The g vector contains the point sources and / or sinks, which are the control or policy variables of the problem, and the boundary conditions of the aquifer system. The Ø vector is a vector of nodal values of the state variable at the grid point, which are the hydraulic head or mass concentration Ø_0 are initial conditions for the problem.

Bredehoeft and Pinder (1973), Huang and Connenfeld (1974), Gupta *et al.* (1975), Young (1977), Grove (1977), Z. Narasimhan *et al.* (1978), Pinder *et al.* (1978) and Gunechten and Gray (1978) solved a large number of groundwater hydaulics and quality problems and concluded the versatality of finite element method to cope up with field problems in groundwater.

An extension of finite differences and finite element digital model can be found in groundwater management models which are also considered as optimization models of the system. These mathematical models consist of a set of economic, hydraulic, water quality or environmental objectives and the management or decision variables that control the groundwater pumping, recharge or waste injection schedules. The alternate management decision are constrained by the groundwater system's hydraulic or water quality response equations and possible well capacity, hydraulic gradient or water demand requirement.

Application of linear programming, stochastic linear programming, quadratic programming, stochastic dynamic programming mixed integer programming and nonlinear programming principles have been used in the last two decades to find optimal solutions for pump location and their spacing, minimising collective draw down, maximising head and minimising total cost of lifting water for meeting certain water demand targets which include industrial, agricultural and municipal demand of water. The timing and staging of well field development, *i.e.*, capacity expansion problem and the design of surface storage and transport facilities to distribute the groundwater supply to the water demand area are also considered in these management models.

In simulation model the response equations are used to predict the response of the groundwater system for a set of pumping or recharge schedules and its hydrogeologic impacts. In contrast optimization models develop optimal planning, design or operational policies for the groundwater system. And because the response equations can be incorporated in the management models – the same equation that would normally be used for simulation – the optimal decisions define not only the optimal pumping and recharge schedules, but also predict time and spatial variation in the groundwater head values. Therefore, management model combines simulation and optimization models and look to the groundwater system in its near totality.

Rosewald and Green (1974), Willis and Newman (1977), Rastogi (1989), Aguado and Remson (1980), Remson and Gorelick (1980), Noel and Howitt (1982), Willis and Finney (1985), Gorelick (1983) and Wanakule *et al.* (1986) have carried out detailed studies on the groundwater simulation management models using linear and non linear programming techniques and have found optimal solutions of the operational or capacity expansion model identifying optimal planning or design policies and simulatenously predicting hydraulic head distribution in the aquifer system. Largely these workers adopted a response equation approach which combines simulation and optimization in a single management model. Though optimal groundwater quality management can be used for ascertaining the feasibility of waste water injection for control of sea water intrusion, in alleviating land subsidence problems or in recharging over developed groundwater basin with reclaimed water, however, computationally these models are very difficult to solve for several reasons. First, the constraint set of the model is nonconvex set. The pumping and injection decision affect the velocity field and, consequently, the convective and dispersive mass transport occurring in the aquifer. Since these terms are multiplied with mass concentration in mass transport equation, the complexity of the problem is very much increased. Further, the size of the constraints set is very large because of temporal discretization of the response equation which affects the optimality of the planning policies. Hence simulation models may be more viable for water quality problems (Willis and Yeh, 1987) compared to optimal management model.

Conclusions

Paper concludes that digital models using finite differences and finite element methods can effectively cope up with complex real world problems involving variable system parameter, temporal and spatial variations in the state and control variables and various boundary conditions. Adequate calibration and validation, however, is a necessary pre-condition for the effective use of a digital groundwater model. Such a model, then, can be used as prediction model, parameter estimation model or a recharge estimation model.

When coupled with optimisation models, digital models can be used as groundwater system management models. These models are primarily concerned with the cost-effectiveness of management schemes which minimize the operational and capital costs for a given planning period for meeting certain water demands.

It must be emphasized that some real life issues are extremely complex and not always exactly quantifiable. Many assumptions and simplifications must be made to construct such models where experience alone decides the data that is important and must be available in required quantity and reliable quality. These systems model actually complement creativity and judgement by carrying out analysis over a wide range and permit to conduct thought experiment to investigate the sensitivity of the model to anticipated changes in aquifer stresses.

It may perhaps be reasonably concluded that system's approach to groundwater resources utilization is under continuous research and development phase at present and therefore, alone, is not enough to directly solve the real world problem.

References

Aquado, E. and Remson, I., (1980) Groundwater management with fixed charges-ASCE Jr. Water Reso. Pin. and Mang. Divn. 106 (WR 2, pp. 375-82.

Bear, J., (1979) Hydraulics of Groundwater-McGraw Hill, New York.

Cooley, (1974) Finite element solutions for the equations of groundwater flow : Desert Research Institute, Univ. of Nevada, Tech. Report Series, H-W, Hydrology and Water Resou. Publ. No. 118, 134p.

Gorelick, S.M., (1983) A review of distributed parameter of groundwater management modelling methods. Water Resou. Res. V. 19, pp. 305-19.

Hall, W.A. and Dracup, J.A., (1970) Water resources systems engineering - McGraw Hill Book Company, New York.

Hayakorn, P.S. and Pinder. G.F., (1983) Computational methods in subsurface flow - Academic Press, New York.

Li W.H., (1972) Differential equations of hydraulic transients dispersion and groundwater flow - Englewood Cliffs, Prentice Hall, NJ.

Maddock, T., III (1972) Algebraic technological function from a simulation model Water Reso. Res., V. 8(1), pp. 129-34.

Noel. J. and Howitt, R., (1982) Conjuctive multibasin management, an optimal control approach. Water Res. Res. V. 18(4), pp. 758-63.

Peaceman, D.W. and Reachfor, H.H., (1955) The numerical solution of parabolic and elliptic differential equations. Soc. Indus. App. Math. Jour. V.3(1), pp. 28-41.

Prickett, T.A., (1975) Modelling techniques for groundwater evaluation. Chow, V.T. Ed., Advances in hydrosciences, V. 10, Academic Press, New York, pp. 1-128.

Rastogi, A.K. (1983) Numerical solutions of confined and unconfined aquifer interactions with partially penetrating rivers. Ph.D. Thesis, Dept. of Civil Engg., Univ. of Birmingham, U.K.

Rastogi, A.K., (1989) Optimal pumping policy and groundwater balance of Blue Lake aquifer-California involving non linear groundwater hydraulics. Jr. Hydrol. V. 111 (1-4), pp. 177-94.

Remson, I. and Gorelick, S.M. (1973) Management models incorporating groundwater variables, in Operation Research in Agriculture and Water Resources, Ed. D. Yaron and C.S. Tapiero, North Holland, Armsterdam.

Remson, J. Hornberger, G.M. and Molz, F.H., (1971) Numerical methods in subsurface hydrology, Wiley Interscience.

Report on Land and Water Development in the Indus Plain, (1964) White House Department of Interior, Panel on Waterlogging and salinity in West Pakistan, The White House, Washington, D.C.

Rosenwald, G.W. and Green, D.W., (1974) A method for determining the optimum location of wells in a reservoir using mixed integer programming. Jr. Petrol. Engg. pp. 44-54.

Rushton. K.R. and Redshaw, S.C., (1979) Seepage and groundwater flow. John Wiley and Sons.

Santing, G., (1963) Groundwater models in the development of groundwater resources with special reference to deltaic areas. Water Reso. Series, UNESCO. 24, pp. 83-84.

Schwarz, J., (1976) Linear models for groundwater management. Jr. Hydrol. V. 28, pp. 377-91.

Smith, G.D., (1979) Numerical solution of partial differential equations. Finite difference methods - 2nd ed. Clarendon Press, Oxford.

Stone, H.K. (1968) Iterative solution of implicit approximations of multidimensional partial differential equations. Soc. Indus. App. Math, Hour, Jr. Numer. Annal. V. 5(3), pp. 530-58.

Trescott, P.C., Pinder, G.F. and Larson, S.J., (1976) Finite difference model for aquifer simulation in two dimension with results of numerical experiments-Techniques of water resources investigation of the USGS, Chapter Cl, Book 7.

Wankule, N., Mays, L.W. and Ladson, L.S., (1986) Optimal management of large scale aquifers : Methodology and application. Water Res. V. 22(4), pp. 447-65.

Weinstein, H.G., Stone, H.L. and Kwan, T.V., (1960) Iterative procedure for solution of systems of parabolic and elliptic equations in three dimensions : Industrial and Engg. chemistry fundamentals, V. 8, No. 2, pp. 2281-87.

Willis, R., and Newman, B.A., (1977) Management model for groundwater development - ASCE, Jr. Water Reso. Pln. and Mang. Divn. 103 (WRI), pp. 159-71.

Willis, R. and Finney, B.A., (1985) Optimal Control of non linear groundwater hydraulic: Thoretical development and numerical experiments. Water Reso. Res. V. 21 (10), pp. 1476-82.

Willis, R. and Yeh, W.W.G., (1987) Groundwater systems planning and management. Prentice Hall International, USA, pp. 82-86.

26

Finite Element Approach for Water Resources Planning

Ashok Kumar Saini, S.R. Singh
and
M.P. Kaushal

Introduction

Planning of water resources is an important aspect for getting higher yield of crops in order to meet the food and fibre requirements of the increasing population of our country. Canal water being almost fully exploited, the increasing demand of irrigation water is met through pumping the groundwater resources from unconfined and confined aquifers thereby lowering the water table at a faster rate. However, in South Western part of Punjab where groundwater is of poor quality and a vast canal net work is used for irrigation, the water level is rising creating water logging problems. Judicious use of the groundwater resource, the response of an aquifer to various stresses need to be studied using modelling techniques. One of the most popular modelling technique is the mathematical models, whereby groundwater flow patterns are obtained mathematically as solution to boundary value problems.

For a compressible porous matrix in two dimensions the equation of continuity in conjunction with Darcy's law yields the following differential equation.

$$\frac{\delta}{\delta x}\left[K\frac{\delta h}{\delta x}\right] + \frac{\delta}{\delta y}\left[K\left(\frac{\delta h}{\delta y} + 1\right)\right] - Q(x, y, t) = \left(C + \frac{S_s \theta}{n}\right)\frac{\delta h}{\delta t} \qquad (1)$$

in which h is pressure head θ is volumetric moisture content, K is hydraulic conductivity, a function of moisture content or pressure head in the zone of aeration and constant at a node in the zone of saturation; Q is sink term; n is porosity; S_s is specific storage coefficient; C is soil moisture capacity, δh/dx; x is space coordinate, positive towards right; y is space coordinate, positive upward; and t is time. Equation (1) is a non linear parabolic partial differential equation. It describes the flow of groundwater in the zone of aeration and unconfined and confined zones of saturation. Eq. (1) for $S_s = 0$ or is commonly known as Richards equation

(Richards, 1931). In saturated porous media C(h) is zero and θ is n.

For many problems of stream aquifer interaction Richards equation or its variant incorporating aquifer compressibility, along with its auxilliary conditions can be written but its analytical solution is very difficult as the eq. is non linear. With the advent of high speed digital computers finite difference and finite element techniques have been applied to solve such problems. Rubin (1968), Freeze (1969, 1971) and Tong and Skaggs (1978) solved Richards eq. in unsaturated saturated domain using finite differences. Neuman (1973) used Galerkins finite element method to solve eq. (1) for problems of seepage through dam and to a drainage ditch. Finite element solution of one dimensional Richards eq. incorporating water uptake by plant roots was developed by Singh & Kumar (1985).

2. General Two Dimensional Problem Formulation

Herein a general two dimensional problem of groundwater flow in saturated unsaturated domain is formulated. It is assumed that a porous medium is resting on an aquitard described by

$$F(x, y) = 0 \tag{2}$$

A porous medium is subjected to infiltration and evaporation at the land surface, a condition very often occurring in the nature. Water is pumped from aquifer at a rate Q(x, y, t) per unit volume of the aquifer per unit time. The two vertical boundaries at the right and left end of the flow domain could be Dirichlet or Neumann boundaries. We assume the origin of coordinate system at the left end of the flow domain (Fig. 26.1) where x axis coincides with the datum and y axis is vertically upwards. The auxilliary conditions for such a general two dimensional problem are as below.

Initial Conditions

$$h(x, y, 0) = h_o(x, y) \tag{3}$$

Boundary Conditions

Let H_{w1} (t) and H_{w2} (t) are the height of water table above datum at x = 0 and L, respectively. We consider the soil moisture to be in equilibrium with the water table, consequently, the pressure head along the two boundaries are as under

$$h(x, y, t)^{x=0} = H_{w1}(t) - y \text{ along STA} \tag{4}$$

$$h(x, y, t)^{x=L} = H_{w2}(t) - y \text{ along RQP} \tag{5}$$

At the land surface the boundary condition becomes

$$-K(h)(\delta h/\delta y + 1)^{y=D} = \text{known rate of infiltration or evaporation} \tag{6}$$

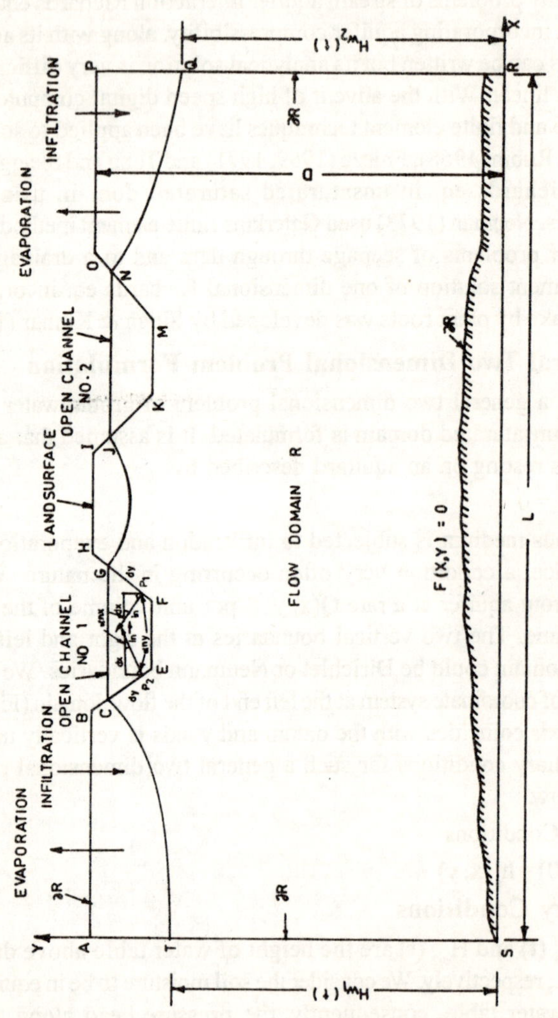

FIG. 16.1 SCHEMATIC OF UNSATURATED SATURATED FLOW DOMAIN.

At the aquitard the normal flux is zero. Hence one gets

$$\frac{(\delta E)}{dx}\frac{(\delta h)}{\delta x}+\frac{(\delta F)}{\delta y}\frac{(\delta h+1)}{\delta y}=0 \quad (7)$$

Differential equation (1) and auxillary conditions (3), (4), (5), (6), (7) constitute a boundary value problem which is solved by finite element technique.

3. Galerkin Finite Element Solution

Eq. (1) is solved using Galerkin's finite element method, the details of which are given by Pinder and Gray (1977). In this approach the approximate solution is represented as a sum of series

$$h(x, y, t) = \sum_{j=1}^{NN} aj(t)\, N_j(x, y) \quad (8)$$

where aj(t) are time dependent coefficient to be determined as a part of solution, N_j (x, y) are basis function and h (x, y, t) is an approximation for h (x, y, t). Application of the method reduces (1) to the following system of nonlinear ordinary differential equation :

$$[H]\{da/dt\} + [G]\{a\} = \{F\} \quad (9)$$

in which [H] and [G] are square matrices of sizes (NN * NN) and {a}, {da/dt} and {F} are column vectors of size (NN*1). Typical elements of these matrices are as below.

$$H_{ij} = \sum_{e=1}^{NE} \int_{R_e} C_e + \theta_e S_s n)\, N_i N_j\, dxdy \quad (10)$$

$$G_{ij} = \sum_{e=1}^{NE} \int_{R_e} K_e \left(\frac{\delta N_i}{dx}\frac{\delta N_j}{dx} + \frac{\delta N_i}{dy}\frac{\delta N_j}{dy}\right) dxdy \quad (11)$$

$$F_{ij} = -\{\sum_{e=1}^{NE} \int_{R_e}^{D} K_e \left(\frac{\delta N}{dy}\right. dx\, dy + \sum_{e=1} \int_{R_e} N_j\, q_n\, ds + \quad (12)$$

$$\sum_{e=1}^{NE} \int_{R_e} Q\, N\, dxdy$$

Eq. (9) constitute a system of first order non-linear ordinary differential equations whose solution will yield aj(t).

Solution of Sytem of differential Equations

Predictor-Corrector Method

Using this method Eq. (9) can be written as below.

Predictor

$$[[H(t)] + (\Delta t/2)[G(t)]]\{a(t+\Delta t/2)\} = [H(t)]\{a(t) + (\Delta t/2)\{F(t)\} \tag{13}$$

Corrrector

$$[[H(t+\Delta t/2)] + (\Delta t/2)[G(t+\Delta t/2)]\{a(t)\} + \Delta t\{F(T+\Delta t/2)\} \tag{14}$$

Fully Implicit and Crank Nicolson Method

In this method fully implicit and Crank Nicolson finite difference analogue of the system of ordinary differential equations (9) is written as below.

$$[[H(t+\Delta t)] + \Delta t\,\text{EPSLON}\,[G(t+\Delta t)]]\{a(t+\Delta t)\} = [[H(t+\Delta t)] - \Delta t(1-\text{EPSLON})[G(t+\Delta t)]]\{a(t)\} + \Delta t\{F(t+\Delta t)\} \tag{15}$$

When EPSLON is equal to one, then it is fully implicit scheme and at EPSLON=0.5 it is Crank Nicolson scheme. In Eq. (15) [H], [G] and {F} are functions of K, C and h which depend upon the unknown pressure head at time $(t+\Delta t)$. To circumvent this problem, we firstly estimate the pressure head at time $(t + \Delta t)$ by extrapolation and then compute K, C and h to estimate values of h. The extrapolation, at any node, is achieved using Taylor series expansion as below.

$$h(t+\Delta t) = h(t) + \Delta t\frac{dh(t)}{dt} + \frac{(\Delta t)^2}{2}\frac{d^2h(t)}{dt^2} + \tag{16}$$

Approximating dh/dt and d^2h/dt^2 by backward finite difference and substituting in (16), one gets :

$$h(t+\Delta t) = 2.5h(t) - 2h(t-\Delta t) + 0.5h(t-2\Delta t) \tag{17}$$

Thus to estimate the pressure head at time $(t+\Delta t)$. One needs its value at three preceeding time steps, t, $(t - \Delta t)$ and $(t-2\Delta t)$. However, for initial two time steps the pressure head are estimated as below.

$$h(\Delta t) = h(0) \quad (18)$$

$$h(2\Delta t) = 2\,h(\Delta t) - h(0) \quad (19)$$

Application of Boundary Conditions

The boudary conditions were applied following the procedure described by Pinder and Gray (1977). Thus application of the boundary condition reduces (13), (14) or (15) to the form :

$$[A]\,\{a\,(t + \Delta t)\} = \{B\} \quad (20)$$

in which [A], {a (t + Δt)} and {B} are reduced matrices of size (NN-m)*(NN-m), (NN-m)*1, and (NN-m)*1, after incorporating the boundary conditions, respectively; and m is the numebr of boundary nodes at which Dirchlet boundary is prescribed.

Van Genuchten's Model

Van Genuchten's (1980) equation is the most flexible model which describe various plausible shapes of moisture characteristic curves by assigning different values to the parameters of the model. His functional form of the model is as below :

$$\theta = \theta_r + \frac{\theta_s - \theta_r}{[1 + \alpha/h/)^n]^m} \quad (21)$$

where θ_r and θ_s are the residual and field saturated volumetric water content, respectively; α and n are empricial constants depending upon soil type and are found from curve fitting of the measured data between h and θ; and m = (1-1/n). Van Genuchten and Nielson (1984) have demonstrated the applicability of this equation for hundreds of soil moisture characteristics of disturbed and undisturbed soils.

Van Genuchten (1980) presented closed form analytic expression for predicting hydraulic conductivity based on his equation of the soil moisture retention data :

$$K(S_e) = K_s \sqrt{S_e}\,[1 - (1 - S_e^{1/m})^m] \quad (22)$$

$$S_e = (\theta - \theta_r) / (\theta_s - \theta_r) \quad (23)$$

in which K_s is the saturated hydraulic conductivity and S_e is the effective fluid saturation.

Validity of the Finite Element Model

The convergence of the Predictor Corrector, Fully Implicit and Crank Nicolson finite difference analogues of the system of nonlinear ordinary differential equation (9), to the author's knowledge, has not been proved yet. Therefore, it is necessary that the finite element - finite difference scheme proposed herein be compared with an analytical solution of the differential equation (1) or its variant, Richards equation (Swartzendurber 1969, p. 223). When gravity and specific storativity are neglected, Eq. (1) reduces to the following form of Richards equation describing groundwater dynamics in a horizontal saturated-unsaturated flow domain :

$$\delta [K(h)\, \delta h/\delta x] / \delta x + \{K(h)\, \delta h/\delta y\}/ \delta y = C(h)\, \delta h/ \delta t \qquad (24)$$

Hornung and Messing (1980) developed an analytical solution of (24) for an artificial flow problem in a square domain, $0 \leq x,\ y \leq 1$. Their solution is as below.

$$h(x, y, t) = \begin{cases} -0.5\,S & \text{if } S \leq 0 \\ -\tan[(\exp(S)-1)/(\exp(S)+1)] & \text{if } S > 0 \end{cases} \qquad (25)$$

where $S = y-x-t$

The properties of the artificial porous medium are :

$$K(h) = \begin{cases} 2/(1+h^2) & \text{if } h < 0 \\ 2 & \text{if } h \geq 0 \end{cases} \qquad (26)$$

$$\theta = \begin{cases} 2[(0.5\pi)^2 - (\arctan h)^2] & \text{if } h < 0 \\ 0.5\,\pi^2 & \text{if } h \geq 0 \end{cases} \qquad (27)$$

$$C(h) = \begin{cases} -4\,(\arctan h)/(1+h^2) & \text{if } h < 0 \\ 0 & \text{if } 0 \geq h \end{cases} \qquad (28)$$

The initial and boundary conditions of this artificial problem are derived from solution eq. (25) by substituting appropriate values of x, y and t. When t = 0 one gets the following initial conditions;

$$h(x, y, 0) = \begin{cases} -0.5\,S1 & \text{if } S1 \leq 0 \\ -\tan[(\exp(S1)-1)/(\exp(S1)+1)] & \text{if } S1 > 0 \end{cases} \qquad (29)$$

where $S1 = y - x$.

Similarly for x = 0 and 1, and y = 0 and 1, one gets the appropriate Dirichlet boundary conditions :

$$h(0, y, t) = \begin{cases} -0.5\,S2 & \text{if } S2 \leq 0 \\ -\tan[(\exp(S2)-1)/(\exp(S2)+1)] & \text{if } S2 > 0 \end{cases} \qquad (30)$$

where $S2 = y - t$,

$$h(x, 0, t) = \begin{cases} -0.5\,S3 & \text{if } S3 \leq 0 \\ -\tan[(\exp(S3)-1)/(\exp(S3)+1)] & \text{if } S3 > 0 \end{cases} \qquad (31)$$

where $S3 = y - 1 - t$,

$$h(x, 0, t) = -0.5\ S4 \qquad \text{if } S4 \leq 0$$
$$-\tan[(\exp(S4)-1)/(\exp(4)+1)] \text{ if } S4 > 0 \qquad (32)$$

where $S4 = y - x - t$,

$$h(x, 1, t) = -0.5\ S5 \qquad \text{if } S5 \leq 0$$
$$-\tan[(\exp(S5)-1)/(\exp(S5)+1)] \text{ if } S5 > 0 \qquad (33)$$

where $S5 = I - x - t$,

The computer programme of the finite element finite difference scheme was suitably modified incorporating the aforesaid initial and boundary conditions to develop numeric solution of eq. (24). The square region of $0 < x, y < 1$ was divided into 16, 24 and 32 (Fig. 26.2) triangular elements. The computations were made for different time interval 0.01, 0.05, 0.10, 0.15, 0.20 and 0.25 day. The analytical as well as numerical solution was obtained and the error was determined using the Tcehbycheff norm.

The numerical solution was obtained by discretizing the space using triangular elements and the solution in time was obtained using Fully implicit, Crank Nicolson and Predictor corrector methods. Error from these methods are given in Tables 26.1, 26.2 and 26.3 respectively.

Table 26.1 using Fully implicit scheme shows that error decreases with passage of time at each time interval for 16 and 24 elements, but it is almost constant in case of 32 elements. The error is minimum when the domain is divided into 24 elements. Table 26.2 using Crank Nicolson Schemes shows that the error is fluctuating like saw tooth for all the elements and at each time interval. In this case the error goes on increasing as the size of time interval increases from 0.01 to 0.25 day. Table 26.3 using Predictor Corrector Scheme also shows that error fluctuates as in case of Crank Nicolson Scheme. Here also error increases as the time interval increases.

Tables 26.1, 26.2 and 26.3 indicate that finite element cum Fully Implicit Scheme gives more accurate results than other two schemes. It is also observed that when small time intervals are used the results are more precise than using higher time intervals.

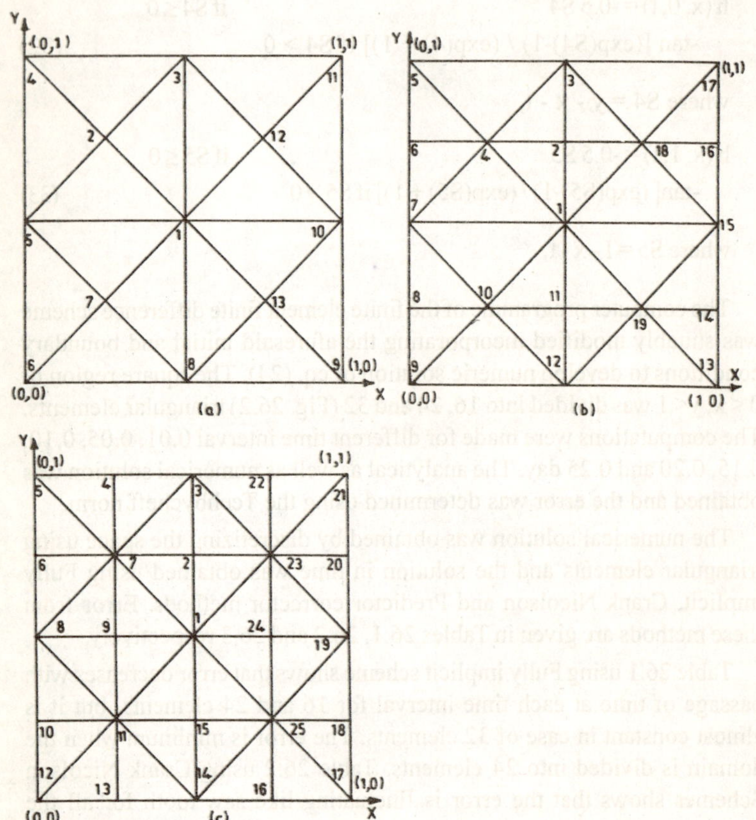

FIG. 26·2 TRIANGULAR FINITE ELEMENT DISCRETISATIONS OF FLOW DOMAIN INTO (a) 16 ELEMENTS (b) 24 ELEMENTS (c) 32 ELEMENTS

Table 26.1 : Error in Finite Element cum Fully implicit scheme

Time interval	*No. of time steps*	*NE=16*	*NE=24*	*NE=32*
0.01 day	5	.2139E-02	.2125E-02	.1524E-01
	10	.1813E-02	.1817E-02	.1532E-01
	15	.1517E-02	.1534E-02	.1537E-01
	20	.1275E-02	.1308E-02	.1540E-01
	25	.1089E-02	.1139E-02	.1540E-01
	30	.9480E-03	.9591E-03	.1540E-01
	35	.8609E-03	.8147E-03	.1544E-01
	40	.8233E-03	.7326E-03	.1545E-01
	45	.8341E-03	.7320E-03	.1544E-01
	50	.8920E-03	.7520E-03	.1541E-01
	55	.7075E-03	.5763E-03	.1542E-01
	60	.5205E-03	.4021E-03	.1543E-01
	65	.3597E-03	.2632E-03	.1543E-01
	70	.2254E-03	.1599E-03	.1544E-01
	75	.1182E-03	.9280E-04	.1544E-01
	80	.3846E-04	.3022E-04	.1544E-01
	85	.1345E-04	.1068E-04	.1544E-01
	90	.3725E-04	.2941E-04	.1544E-01
	95	.3279E-04	.2586E-04	.1544E-01
	100	.5960E-04	.2586E-04	.1544E-01
0.05	5	.1022E-02	.9921E-03	.1543E-01
	10	.7841E-03	.6271E-03	.1543E-01
	15	.1183E-03	.9289E-04	.1544E-01
	20	.5960E-03	.2980E-07	.1544E-01
0.10	1	.1592E-02	.1365E-02	.1533E-01
	2	.7841E-03	.6271E-03	.1543E-01
	3	.7949E-03	.6864E-03	.1546E-01
	4	.6286E-03	.5101E-03	.1546E-01
	5	.8584E-03	.4830E-03	.1545E-01
	6	.5190E-03	.4004E-03	.1543E-01
	7	.2256E-03	.1602E-03	.1544E-01
	8	.3862E-04	.3026E-04	.1544E-01
	9	.3716E-04	.2938E-04	.1544E-01
	10	.5960E-07	.2980E-07	.1544E-01

Table 26.1(Contd.)

Time interval	*No. of time steps*	*NE=16*	*NE=24*	*NE=32*
0.15	1	.1294E-02	.9692E-02	.1537E-01
	2	.7344E-03	.5052E-03	.1549E-01
	3	.5155E-03	.3470E-03	.1547E-01
	4	.3873E-03	.2467E-03	.1545E-01
	5	.1181E-03	.9277E-04	.1544E-01
	6	.3713E-04	.2932E-04	.1544E-01
	7	.1192E-06	.5960E-07	.1544E-01
0.20	1	.1011E-02	.5943E-03	.1542E-01
	2	.4658E-03	.2180E-03	.1550E-01
	3	.2617E-03	.1017E-03	.1547E-01
	4	.3837E-04	.2999E-04	.1544E-01
	5	.5960E-07	.2980E-07	.1544E-01
0.25	1	.7579E-03	.3128E-03	.1544E-01
	2	.3454E-03	.1260E-03	.1544E-01
	3	.1177E-03	.9247E-04	.1544E-01
	4	.5960E-07	.2980E-07	.1544E-01

Table 26.2 : Error in Finite Element cum Crank Nicolson Scheme

Time interval	*No. of time steps*	*NE=16*	*NE=24*	*NE=32*
0.01 day	5	.4666E-02	.4681E-02	.2998E-01
	10	.4276E-02	.4284E-02	.5921E-01
	15	.4475E-02	.4421E-02	.2719E-01
	20	.3758E-02	.3803E-02	.5931E-01
	25	.4359E-02	.4349E-02	.2722E-01
	30	.3445E-02	.3451E-02	.5872E-02
	35	.4328E-02	.4305E-02	.2727E-01
	40	3332E-02	.3240E-02	.5821E-02
	45	.4385E-02	.4336E-02	.2727E-01
	50	.3414E-02	.3775E-02	.5909E-02
	55	.4419E-02	.4361E-02	.2725E-01
	60	.3003E-02	.2886E-02	.5872E-02
	65	.4394E-02	.4344E-02	.2727E-01
	70	.2713E-02	.2651E-02	.5854E-02

Table 26.2 (Contd.)

Time interval	*No. of time steps*	*NE=16*	*NE=24*	*NE=32*
	75	.4377E-02	.4334E-02	.2727E-01
	80	.2549E-02	.2532E-02	.5847E-02
	85	.4367E-02	.4329E-02	.2727E-01
	90	.2543E-02	.2507E-02	.5843E-02
	95	.4367E-02	.4328E-02	.2727E-01
	100	.2547E-02	.2516E-02	.5845E-02
0.05 day	5	.2232E-01	.2253E-01	.4422E-01
	10	.1395E-01	.1322E-01	.1413E-01
	15	.2209E-01	.2230E-01	.4386E-01
	20	.1388E-01	.1254E-01	.1347E-01
0.10 day	1	.4914E-01	.4908E-01	.7174E-01
	2	.9281E-02	.1078E-01	.1262E-01
	3	.4760E-01	.4781E-01	.6997E-01
	4	.1308E-01	.1445E-01	.1623E-01
	5	.4740E-01	.4763E-01	.6966E-01
	6	.1366E-01	.1504E-01	.1685E-01
	7	.4731E-01	.4757E-01	.6960E-01
	8	.1371E-01	.1510E-01	.1693E-01
	9	.4729E-01	.4756E-01	.6959E-01
	10	.1387E-01	.1512E-01	.1700E-01
0.15 day	1	.7409E-01	.7400E-01	.9661E-01
	2	.8403E-02	.1005E-01	.1196E-01
	3	.7316E-01	.7322E-01	.9550E-01
	4	.1066E-01	.1238E-01	.1436E-01
	5	.7304E-01	.7315E-01	.9542E-01
	6	.1081E-01	.1248E-01	.1442E-01
	7	.7303E-01	.7314E-01	.9542E-01
0.20 day	1	.9909E-01	.9897E-01	.1215
	2	.7421E-02	.9166E-02	.1109E-01
	3	.9864E-01	.9859E-01	.1209
	4	.8231E-02	.1071E-01	.1230E-01
	5	.9861E-01	.9857E-01	.1209
0.25 day	1	.1241	.1239	.1464
	2	.6544E-02	.8436E-02	.1045E-01
	3	.1239	.1238	.1463
	4	.6721E-02	.8565E-02	.1053E-01

Table 26.3 : Error in Finite Element cum Predictor-Corrector Scheme

Time interval	*No. of time steps*	*NE=16*	*NE=24*	*NE=32*
0.01 day	5	.4753E-02	.4697E-02	.2698E-01
	10	.4316E-02	.4301E-02	.5919E-02
	15	.4497E-02	.4421E-02	.2719E-01
	20	.3783E-02	.3822E-02	.5936E-02
	25	.4366E-02	.4350E-02	.2721E-01
	30	.3458E-02	.3462E-02	.5872E-02
	35	.4327E-02	.4306E-02	.2727E-01
	40	.3343E-02	.3254E-02	.5823E-02
	45	.4387E-02	.4339E-02	.2726E-01
	50	.3423E-02	.3283E-02	.5909E-02
	55	.4420E-02	.4360E-02	.2725E-01
	60	.3006E-02	.2887E-02	.5871E-02
	65	.4394E-02	.4344E-02	.2726E-01
	70	.2714E-02	.2651E-02	.5853E-02
	75	.4376E-02	.4334E-02	.2727E-01
	80	.2535E-02	.2525E-02	.5846E-02
	85	.4366E-02	.4329E-02	.2727E-01
	90	.2514E-02	.2511E-02	.5842E-02
	95	.4365E-02	.4328E-02	.2727E-01
	100	.2527E-02	.2520E-02	.5845E-02
0.05 day	5	.2203E-01	.2255E-01	.4443E-01
	10	.1320E-01	.1312E-02	.1402E-02
	15	.2175E-01	.2229E-01	.4401E-01
	20	.1300E-01	.1247E-02	.1338E-02

References

Freeze, R.A., (1969) The Mechanism of Natural Groundwater Recharge and Discharge, 1, One Dimensional Vertical, Unsteady Unsaturated Flow above a Recharging or Discharging Groundwater Flow System, Water Resources Research V. 5, pp. 153.

Freeze, R.A., (1971) Three Dimensional, Transient, Saturated-Unsaturated Flow in a Groundwater Basin, Water Resources Research, V. 7(2), pp. 347-66.

Hornung, U. and Messing, W., (1980) A Predictor-Corrector Alternating

Direction Implicit Method for Two-Dimensional Unsteady Saturated-Unsaturated Flow in Porous Media. J. Hydrol. V. 47, pp. 317-23.

Neuman, S.P., (1973) Unsaturated-Saturated Seepage by Finite Elements. J. of Hydraulic Div. ASCE, 99 (HY-12) pp. 2233-90.

Pinder, G.F. and Gray, W.C., (1977) Finite Element Simulation in Surface and Subsurface Hydrology, Academic Press, New York.

Richards, L.A., (1931) Capillary Conduction of Liquids Through Porous Medium, physics, V., pp. 318-31.

Rubin, J., (1968) Theoretical Analysis of two Dimensional Transient Flow of Water in Unsaturated and Partly Unsaturated Soils. Soil Sci. Soc. Amer. Proc. V. 32(5), pp. 607-09.

Singh, S.R. and Kumar, A.K., (1985) Analysis of Soil Water Dynamics: II. A Finite Element Model for Aquifer-Soil Plant Atmosphere Interactions in Irrigated Lands. J. Agric. Engg., ISAE, V. 22(2), pp. 50-74.

Swartzendruber, D., (1989) The Flow of water in Unsaturated Soil in Flow Through Porous Media, ed. R.J.M. De Wiest Academic Press, NY.

Tong, Y.K. and Skagges, R.W., (1978) Subsurface Drainage in Soil with High Hydraulic Conductivity Layers, Trans. of the ASAE, V. 21(3), pp. 515-21.

Van Genuchten, M. Th., (1980) A closed form equation for Predicting the Hydraulic Conductivity of Unsaturated Soils. Soil Sci. Soc. Am. J. V. 44(5), pp. 892-98.

27

Flood Simulation in Adyar River System

S. Mohan, H. Raman
and
K. Srinivasa Raju

Introduction

Floods are caused due to vagaries of the monsoon in our country and results in considerable damage to property and lives and eradication of communications over large areas. The results of flood damage may be of direct or indirect nature. To face this complex situation, there must be a serious plan of implementing flood control management strategies which are inevitable in the present situation.

Simulation models have been used in the past to solve multi-objective operational problems efficiently. The study by Harvard water resources group (Maass *et al.* 1962) shows that system analysis approach using simulation is the only tool that would provide designs and operating policies. Significant studies that derive operating policies begin with the work of Haf Schmidt and Fiering (1966) who developed programmes which could deal with a large number of alternative designs in many different combinations for economic analysis. Sigvaldason (1976) introduced a mathematical model for assessing alternative policies of operation for the Trent river system consisting of 48 reservoirs in Ontario, Canada. The reservoir system is used for flood control, water supply, hydropower and augmentation of flows through the canal system during summer period. Srivastava and Banday (1980) presented an operation study of two proposed detention basins in the River Jhelum in Kashmir Valley and developed three flood hydrographs to study the basin behaviour during floods. Sengupta *et al.* (1982) suggested a dynamic model for hourly forecasting of river flow during storm by learning identification algorithm, correlating lagged instant flows at the different upstream gauging stations and the model is verified by field measurements at Domohani bridge on the river Teesta in North Bengal. Colon and McMahon (1987) developed BRASS (Basin Runoff and Streamflow Simulation Model) to improve the real-time and predictive determination of flood discharges and stages and to aid in flood management decisions with Savannah river system.

Shane and Gilbert (1982) developed a hydrosystem simulation model, HYDROSIM, for the TVA reservoir management personnel for evaluation of the impact of new operating requirements on established objectives like forecasting of reservoir system operation. HEC 3 model (1974) intended for reservoir systems analysis for conservation has been proved to be the best simulation model among the available reservoir simulation models. It can do multi-purpose multi-reservoir simulation and capable in handling thirty reservoirs. It can also accept any non-looping configuration of reservoirs. HEC 5 model (1979) (simulation of flood and conservation systems) can perform monthly daily or hourly operations of multi-purpose reservoir systems. Shnaidman (1989) used an aggregative simulation approach in the simulation modelling of Terck River Basin. The experiments showed the usefulness of this approach due to the block principle of development of the model and verification of the description of individual elements.

In this study, an hourly simulation model is developed and applied to Adyar River System, consisting of five major tanks, namely, Neman, Sriperumpudur, Chembarambakkam, Manimangalam and Pillaipakkam in Tamil Nadu. By using the proposed methodology flood control strategy was also derived.

System Description

Adyar River has its origin at surplus course of Manimangalam tank on western side. The length of the river is 40 Km. of which 10 Km. length flows through south Madras city and joins Bay of Bengal. The catchment area of this river basin is 857 sq. km. The river has two arms: the northern arm comes from Chembarambakkam minor basin and joins the southern arm comes from Guduvancheri at Tiruneermalai; the southern arm has no storage structure since the flat topography does not permit the same. The capacities of the five Tanks in the Adyar River system are 7.3, 4.9, 88.3, 3.5 and 6.0 Mcum. respectively. Among these tanks Chembarambakkam is the largest tank serving both for flood control and irrigation. The surplus flows from Nemam and Sriperumpudur tanks find its way to the Chembarambakkam tank and from these the water releases to the Adyar river are controlled by surplus weirs and a regulator. The tanks at Pillaipakkam and Manimangalam flow surplus directly into the Adyar river. There is also a significant contribution by the overland flow into the right arm of the Adyar river. The Chembarambakkam tank has a catchment area of 354.2 sq. km. and a command area of 5269 ha. The main crop is paddy and the other crops raised in the basin are sugarcane, banana, betelvine, groundnut. The average annual rainfall in this region is around 1200 mm, with 65% contribution of rainfall from north-east monsoon. Evaporation rates are varying from 3 to 8 mm. per day. The schematic diagram of Adyar river system is shown in Fig. 27.1.

The Adyar river has been experiencing floods in 1930, 1937, 1943, 1960, 1966, 1976 and 1985. The damage was severe with respect to the human loss, private and public property, worth several crores of rupees. In 1976 flood, a low pressure was located on 22nd of November in the Bay about 1000 Km. south-east of Madras. The rainfall was as high as around 450 mm. The recurrence interval for this one day rainfall is 1000 years. The peak discharge was 2067 cumecs. Chembarambakkam tank has also contributed considerably to the floods in Adyar river. In 1985 flood, a low pressure formed in the southern Bay of Bengal and intensified into a depression. There was a heavy rainfall for three continuous days (11th to 14th of November). The rainfall was around 300 mm. The recurrence interval of the maximum three day rainfall is 300 years and the peak discharge during this flood was 2040 cumecs. The Chembarambakkam arm of the river did not contribute in this flood. The computed flood levels during 1976 and 1985 floods are shown in Fig. 27.2.

Methodology

Simulation of inflows were carried out for five major tank systems in Adyar river. It was observed that there was considerable overflows from Neman and Sriperumpudur tanks to Chembarambakkam tank. Manimangalam and Pillaipakkam tanks give away their surplus to Adyar river directly along with overland flow. The methodology followed in this study is explained below.

Step 1 : The inflows into Nemam, Sriperumpudur, Pillaipakkam and Manimangalam tanks were simulated sequentially using the developed hourly simulation model.

Step 2 : The inflows into Chembarambakkam tank (upstream releases from Nema and Sriperumpudur tanks plus local flows of Chembarambakkam tank) are used as input to simulate the overflows from Chembrambakkam tank.

Step 3 : The overland flow contribution from the right arm of the Adyar river was computed using a rainfall runoff model.

Step 4 : The irrigation releases are checked with the requirements; the total overflows from Chembarambakkam, Pillaipakkam and Manimangalam tanks and overland flow are checked for feasibility of accommodating flood flows into the Adyar river.

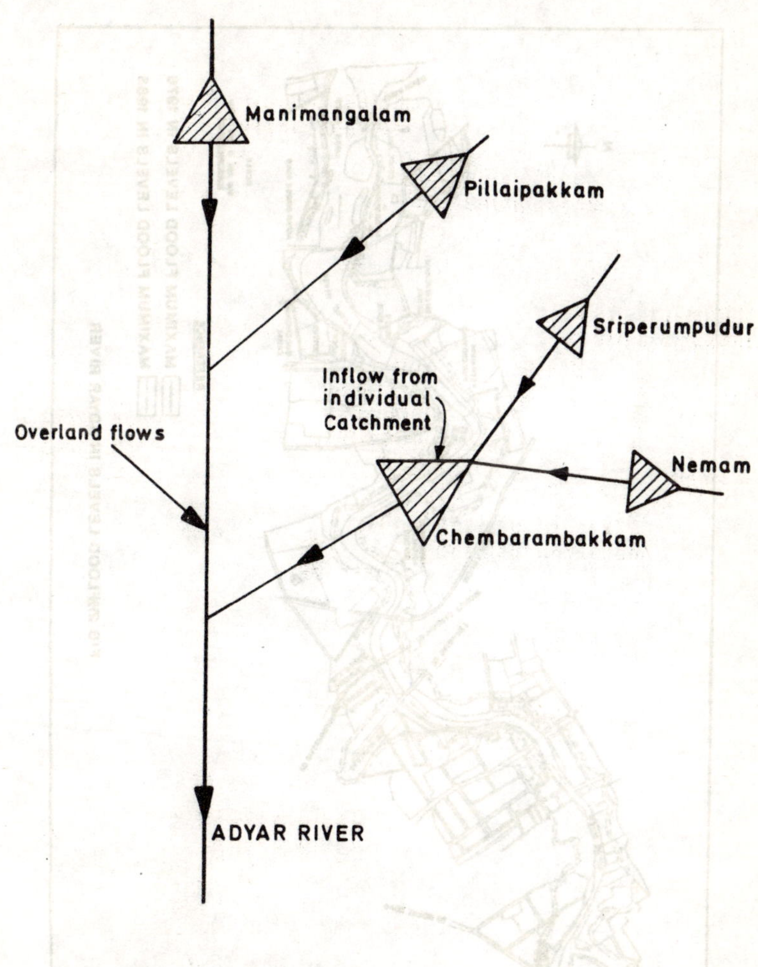

FIG. 27-1 SCHEMATIC DIAGRAM OF ADYAR RIVER SYSTEM.

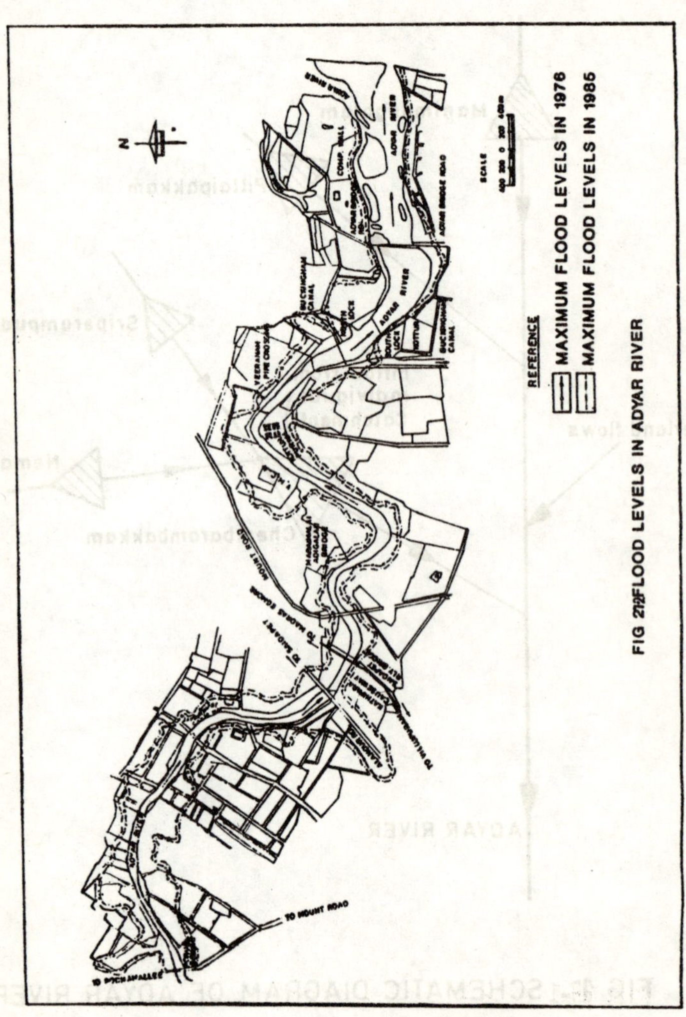

FIG 212 FLOOD LEVELS IN ADYAR RIVER

Step 5 : If the total flow is more than the flood capacity of river then flood control management strategy is derived by suitably increasing the capacity of Chembarambakkam, Pillaipakkam and Manimangalam tanks.

Computations in the model were based on the following equation of continuity.

$$S_{t+1} = S_t + Q_t - R_t - E_t - O_t$$

where Q_t, R_t, E_t, O_t are inflow, release, evaporation, overflow (all in Mcum) respectively in a period. S_t and S_{t+1} are initial and final storages in Mcum for the period t. The details of the developed programme are as follows.

With the known initial storage and computed final storage, average storage was calculated for the simulation period. Based on the area-capacity curves, water spread area corresponds to average storage was found. With the known evaporation rate, loss due to evaporation (Mcum) was computed and using continuity equation final storage was recomputed. If the storage at the end of the period is less than zero, it is equated to zero. If storage is greater than the capacity, overflow is calculated and storage is equated to the capacity of the tank. Again average storage was calculated on the basis of newly obtained final storage. If the difference between final storages between two consecutive steps is less than 0.01 *Mcum, then the programme is taken up to the next period and above computations were repeated for all periods. The main input for the programme are inflows, capacity, initial storage, water spread area, irrigation demands, evaporation rate and the output includes irrigation release, overflows, evaporation and final storage.

Analysis

The hourly rainfall data (mm) used in this study was collected from the Indian Meterological Department for the periods November 1976 and November 1985, with the accuracy of half a millimetre. Hourly evaporation rate (mm/hour) estimated for Red hills lake, located nearby Madras, is used in this study. Rational formulae are used to calculate runoff from rainfall data. A variable runoff co-efficient in the range of 0.25-0.40 suggested by Mohankrishnan (1989) was adopted in this study. The developed methodology has been applied for multi-reservoir system for simulating November, 1976 and November 1985, floods.

Figure 27.3 shows computed flood hydrograph for the period November 24-25, 1976. As the observed values for this flood were not available, the comparison was done only with respect to flood peak value. The prominent peaks one at 21st hour and the other at 28th hour were observed from this

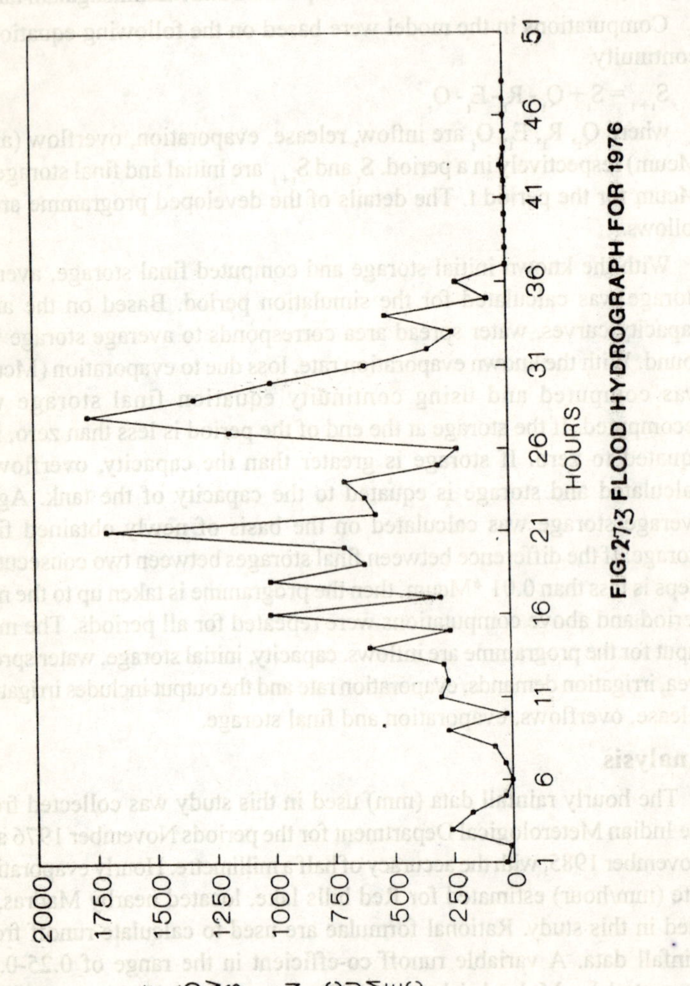

FIG.27.3 FLOOD HYDROGRAPH FOR 1976

figure and the magnitude of highest peak was 1774 cumecs. This is in good comparison with the actual observed peak with a magnitude of 2,067 cumecs. It was observed from the results that the overflows were appreciably high between 31st and 36th hours.

Figure 27.4 shows the comparison of actual and computed flood hydrographs during 1985. It can be seen that the flood duration was more than 1976 flood. The computed maximum flood magnitude was 1833 cumecs which is comparable with actual observed flow of 2040 cumecs. The overflow contribution from Chembarambakkam tank was reduced to zero due to the increased capacity by the construction of bund during the year 1978.

It was reported that maximum flood flow carrying capacity of Adyar river is 800 cumecs. Several simulation runs were made to derive the control strategy to reduce the flood flow into Adyar river.

From this study it was found that the capacity of Chembarambakkam tank should be increased from 88.3 to 93 Mcum, such that the flood flow into the Adyar river from left arm would be reduced to its minimum. Similar suggestion was also made earlier (Expert Committee, 1986). From 1985 flood studies it was noticed that there was a significant overflow of water from Manimangalam and Pillaipakkam tanks to Adyar river. From this study it is deduced that by increasing the storage capacities of these tanks (8.0 and 6.0 Mcum) the flood hazard in Adyar river system would be minimised.

References

Colon, R. and McMahon, G.V., (1987) BRASS Model : Application to Savannah River System. J Water Reso. Plann. & Manag. Div., ASCEV. 113, pp. 177-90.

Expert Committee Report on Flood Protective Works in Adyar River (1986) Public Works Department Publication, Govt. of Tamil Nadu, Madras.

Hufschmidt, M.M. and Fiering, M.B., (1966) Simulation Technique for Design of Water Resources Systems. Harvard University Press, Cambridge.

Maass, A. and Hufschmidt. N.M. (1962) Methods and Techniques of Analysis of Multi Unit Multi Purpose Water Resources Systems : A general statement, Design of Water Resources Systems : A general statement, Design of Water Resources Systems. ed. Maass *et al*. Harvard University Press, Cambridge, pp. 247-62.

Mohanakrishan, A., (1989) Runoff Assessment in the River Basins of Tamil Nadu. Lecture notes on Computer Aided Statistical Analysis in

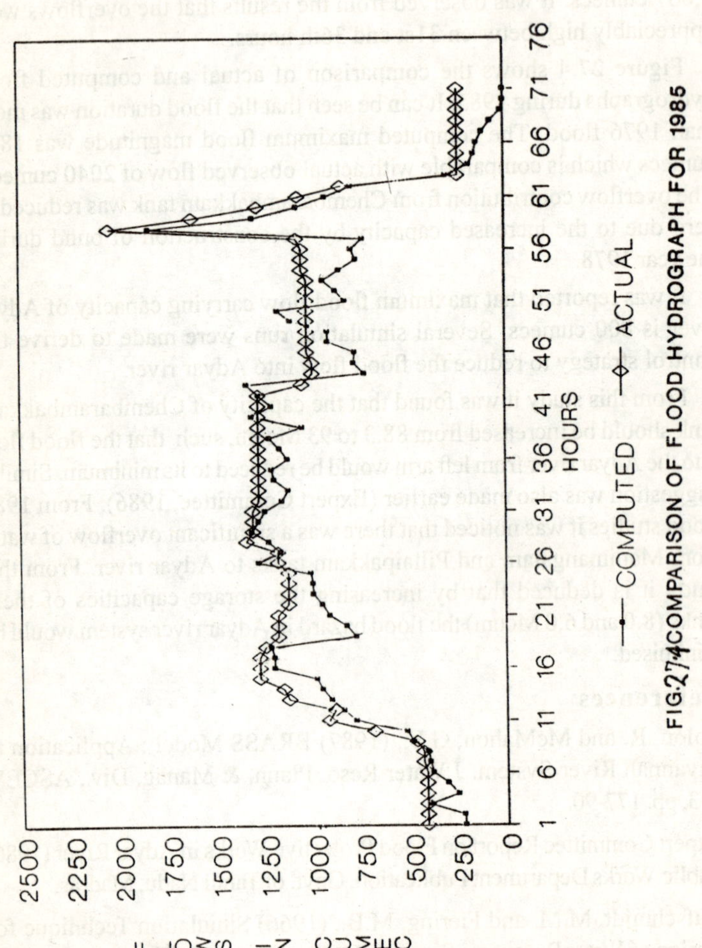

FIG:27.4 COMPARISON OF FLOOD HYDROGRAPH FOR 1985

Hydrology and Water Resources Engineering. Dept. of Civil Engineering, Indian Instt. of Technology, Madras pp. 3601-09.

Report on Floods in Madras (1976) Public Works Department Publication, Government of Tamil Nadu, Madras.

Reservoir System Analysis for Conservation (HEC 3) (1974), User manual, Hydrologic Engineering Centre, U.S. Army Corps of Engineers, Davis.

Reservoir System Operation for Flood Control and Conservation (HEC 5) (1979), User manual. Hydrologic Engineering Centre, U.S. Army Corps of Engineers Davis.

Sengupta, S.K., Maulik, M.N. and Chaudhuri, A.S., (1982) Forecasting of river Flow during storm periods by a learning identification algorithm. J. of Inst. of Engi. (India) Civil Engineering Division, V. 63, pp. 6-9.

Shane, R.M. and Gilbert, K.C., (1982) TVA Hydro Scheduling Model : Practical aspects. J. of Water Resou. Plann. & Manage. Div., ASCE V. 108, pp. 1-19.

Shnaidman, V.M., (1989) Water Releases Calculations by means of an aggregative simulation model. Water Resou. Res. V. 15, pp. 277-84.

Sigvaldason, O.T., (1976) A simulation model for operating a multipurpose multireservoir system, Water Resou. Res. V. 12, pp. 263-76.

Srivastava, D.K. and Banday, Z.A., (1980) Operation study of flood detention basins in River Jhelum in Kashmir Valley, J. of Inst. of Eng. (India) Civil Engineering Division, V. 61, pp. 61-67.

28

Numerical Modeling for the Estimation of Recharge in Tiruchirapalli Aquifers, Tamil Nadu

S. SATHYAMOORTHY, N. DHAKSHNAMOORTHY
AND
A. BALASUBRAMANIAN

Introduction

Groundwater modeling is a methodology for the analysis of mechanisms and controls of groundwater systems and for the evaluation of policies, as actions and designs that may affect such systems. Modelling serves as a means to ensure orderly interpretation of data describing a groundwater system, and to ensure that this interpretation is a consistent representation of the system. It can also provide a quantitative indicator for resource evaluation where the available field data are limited. It will be also possible to impose on a model future conditions which may develop such as increased or redistributed pumping, increased or decreased recharge due to engineering structures, change in recharge patterns and many other factors and analyse the expected response for the proper planning and decision making. In short, groundwater models can be used as predictive as well as deterministic tools for a sound groundwater management. Balasubramanian and Sastri (1989) have developed the two dimensional finite difference model of Tambraparni river basin, Tamil Nadu in order to predict the effects of proposed pumping.

An attempt has been made in this study to : (i) simulate the groundwater system around Tiruchirapalli in Tamil Nadu by constructing and operating a mathematical model which assumes the appearance of the actual aquifer behaviour and (ii) evaluate the groundwater available in storage and due to recharge under the existing conditions.

Area of Study

Tiruchirapalli is the fourth largest town in Tamil Nadu and lies at a distance of 317 km. from Madras by road. Falling mostly within parts of Tiruchirapalli and Lalgudi taluks of Tiruchirappali district (Fig. 28.1). The study area extends over approximately 400 sq. km. and is bounded between latitudes 10° 34′19″ and 10°55′00″ N and longitudes 78°34′19″ and

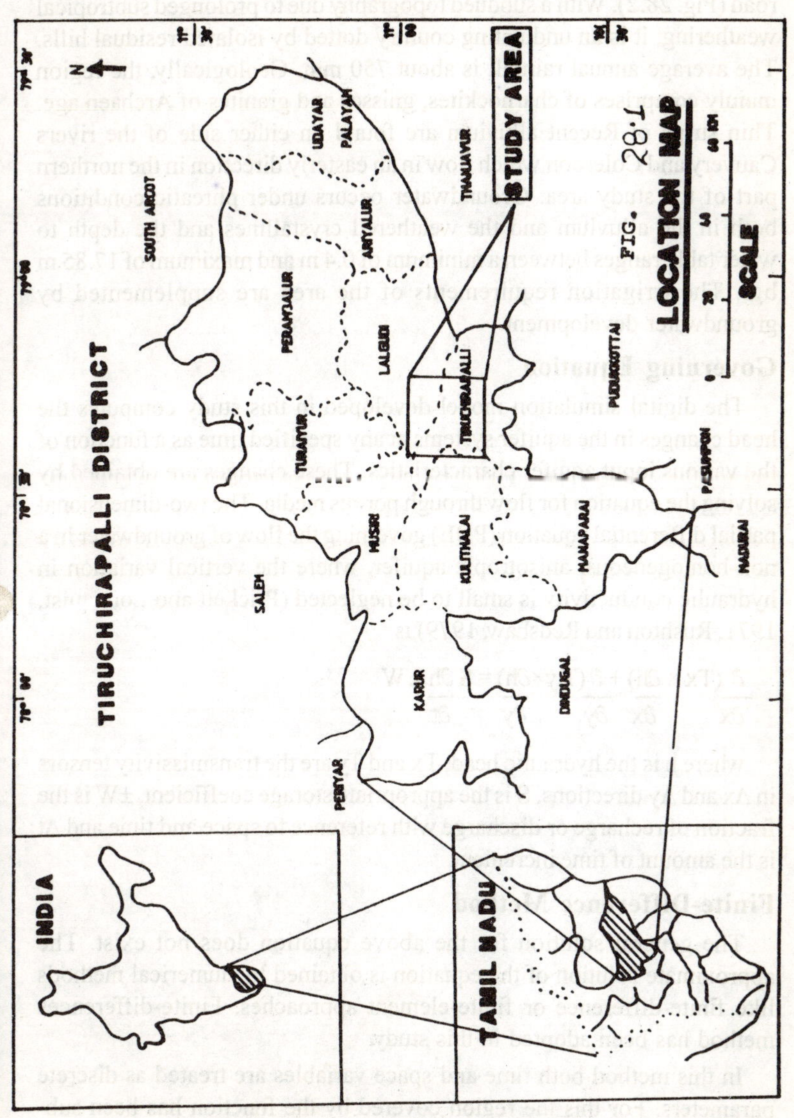
TIRUCHIRAPALLI DISTRICT
LOCATION MAP
STUDY AREA
SCALE
INDIA
TAMIL NADU
SOUTH ARCOT
PERAMBALUR
UDAYAR PALAYAM
ARIYALUR
THANJAVUR
LALGUDI
TIRUCHIRAPALLI
PUDUKKOTTAI
THURAIYUR
MUSIRI
SALEM
KULITHALAI
MANAPARAI
KARUR
DINDIGUL
PERIYAR

FIG. 28.1

78°45′00″ E. This area is easily approachable through National-State Highways and Southern Railways. There is an international airport located at a distance of 6 km from Tiruchirapalli railway junction on the Pudukkottai road (Fig. 28.2). With a subdued topography due to prolonged subtropical weathering, it is an undulating country dotted by isolated residual hills. The average annual rainfall is about 750 mm. Geologically, the region mainly comprises of charnockites, gnisses and granites of Archaen age. Thin strips of Recent alluvium are found on either side of the rivers Cauvery and Coleroon which flow in an easterly direction in the northern part of the study area. Groundwater occurs under phreatic conditions both in the alluvium and the weathered crystallines and the depth to water table ranges between a minimum of 0.4 m and maximum of 17.85 m bgl. The irrigation requirements of the area are supplemented by groundwater development.

Governing Equation

The digital simulation model developed in this study computes the head changes in the aquifer systems at any specified time as a function of the various input aquifer characteristics. These changes are obtained by solving the equation for flow through porous media. The two-dimensional partial differential equation (PDE) governing the flow of groundwater in a non-homogeneous, anisotropic aquifer, where the vertical variation in hydraulic conductivity is small to be neglected (Prickett and Lonnquist, 1971, Rushton and Redshaw, 1979) is

$$\frac{\partial}{\partial x}\left(Tx \times \frac{\partial h}{\partial x}\right) + \frac{\partial}{\partial y}\left(Ty \times \frac{\partial h}{\partial y}\right) = S\frac{\partial h}{\partial t} \pm W$$

where h is the hydraulic head, Tx and Ty are the transmissivity tensors in Δx and Δy directions, S is the appropriate storage coefficient, $\pm$W is the fraction of recharge or discharge with reference to space and time and Δt is the amount of time increment.

Finite-Difference Method

The general solution for the above equation does not exist. The approximate solution of the equation is obtained by numerical methods like finite-difference or finite-element approaches. Finite-differences method has been adopted in this study.

In this method both time and space variables are treated as discrete parameters. For this the region covered by the function has been sub-divided into 13 columns and 14 rows amounting to 182 nodes in which 50 are boundary nodes and 132 nodes lying within these boundary nodes

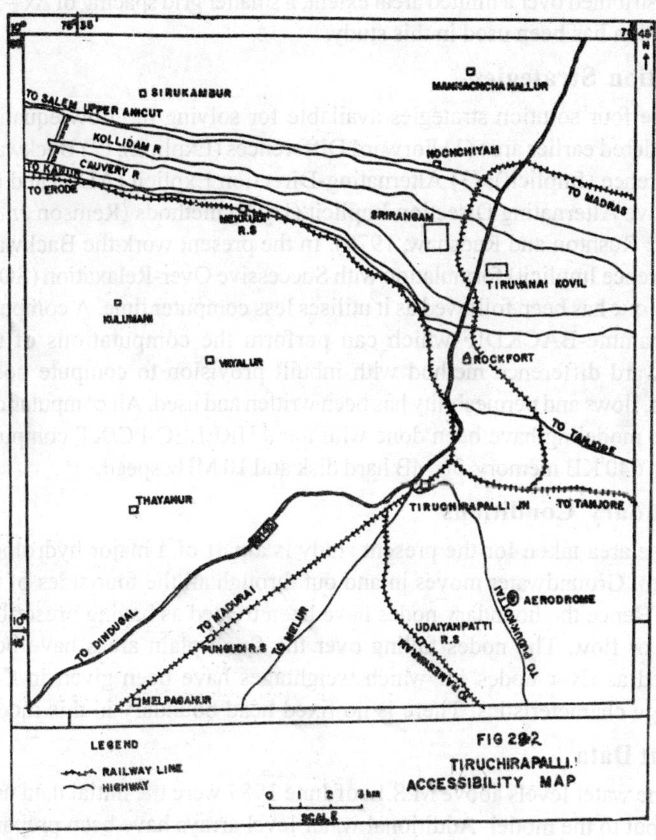

FIG 29.2
TIRUCHIRAPALLI
ACCESSIBILITY MAP

are the active nodes of the model (Fig. 28.3). As the aquifers under study are distributed over a limited areal extent, a smaller grid spacing of $\Delta x = \Delta y = 1500$ m has been used in this study.

Solution Strategies

The four solution strategies available for solving the flow equation considered earlier are : (1) Forward Differences (Explicit), (2) Backward Difference (Implicit), (3) Alternating-Direction Explicit (ADE) and (4) Iterative Alternating Direction Implicit (IADI) methods (Remson *et al.*, 1971.; Rushton and Redshaw, 1979). In the present work the Backward Difference Implicit) formulation with Successive Over-Relaxation (SOR) technique has been followed as it utilises less computer time. A computer Programme-BACKDIF-which can perform the computations of the backward difference method with inbuilt provision to compute nodal yields, flows and permeability has been written and used. All computations of the modeling have been done with the AURELEC-PC/XT computer with a 640 KB memory, 30 MB hard disk and 10 MHz speed.

Boundary Conditions

The area taken for the present study is a part of a major hydrologic system. Groundwater moves in and out through all the four sides of the area. Hence the boundary nodes have been treated as having prescribed head or flow. The nodes falling over the flood plain areas have been treated as river nodes for which weightages have been given in their storage characteristics. There is no fixed head boundary in this model.

Input Data

The water levels above M.S.L. of June 1987 were the initial data used as input to the model. Additional water level arrays have been prepared in order to calculate the discharge and recharge volumes. Initial nodal transmissivity values (Fig. 28.4) of the model have been taken from the results of the dugwell pumptests. Though the storage coefficient values of pumping tests had a wide range, experimental runs have shown that a value of 0.1 for the alluvial aquifers and 0.01 to 0.03 for hardrock aquifers are most suited. The aquifer basement topography (Fig. 28.5) has been evaluated from the analysis and interpretation of geophysical soundings. Taking the change in the saturated thickness of the aquifer from June '87 to December '87 and the storage coefficient, aquifer inflows from precipitation and other sources have been computed for all the nodes. This array has been kept as a basis for all types of simulations made by increasing or decreasing the inflows-outflows depending upon the

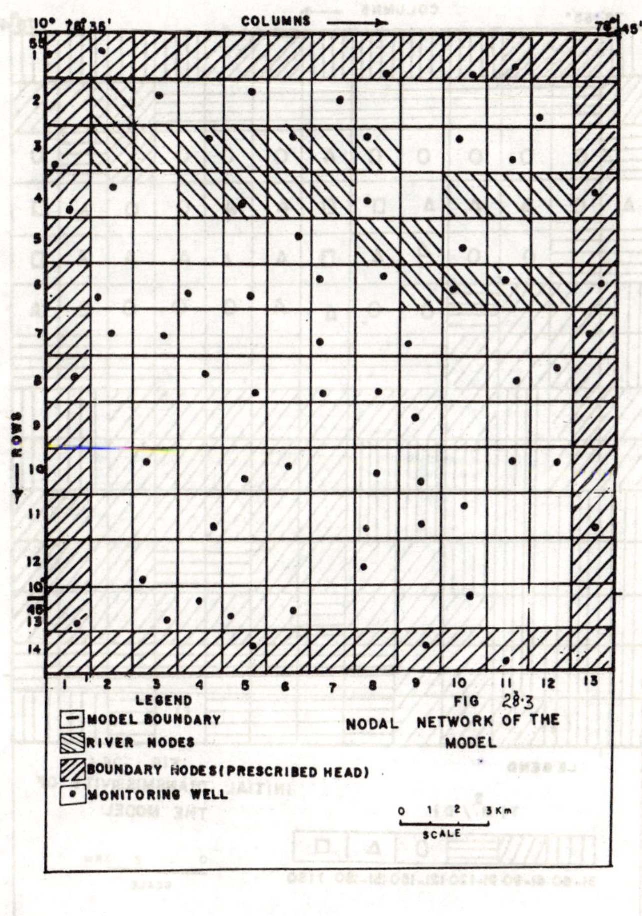

FIG 28.3
NODAL NETWORK OF THE MODEL

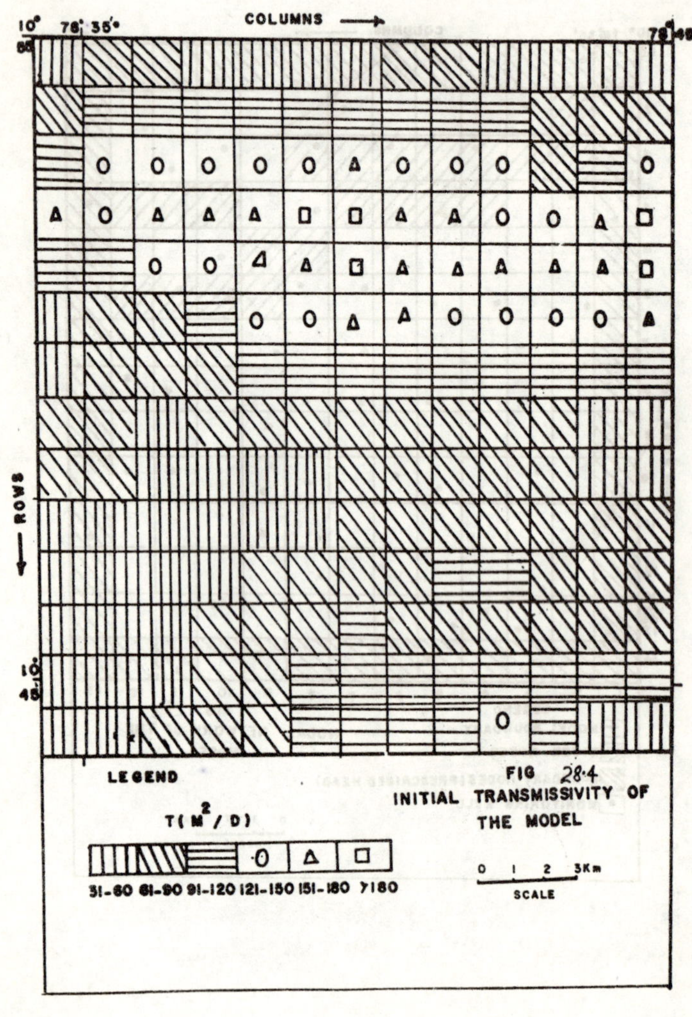

FIG 28.4
INITIAL TRANSMISSIVITY OF THE MODEL

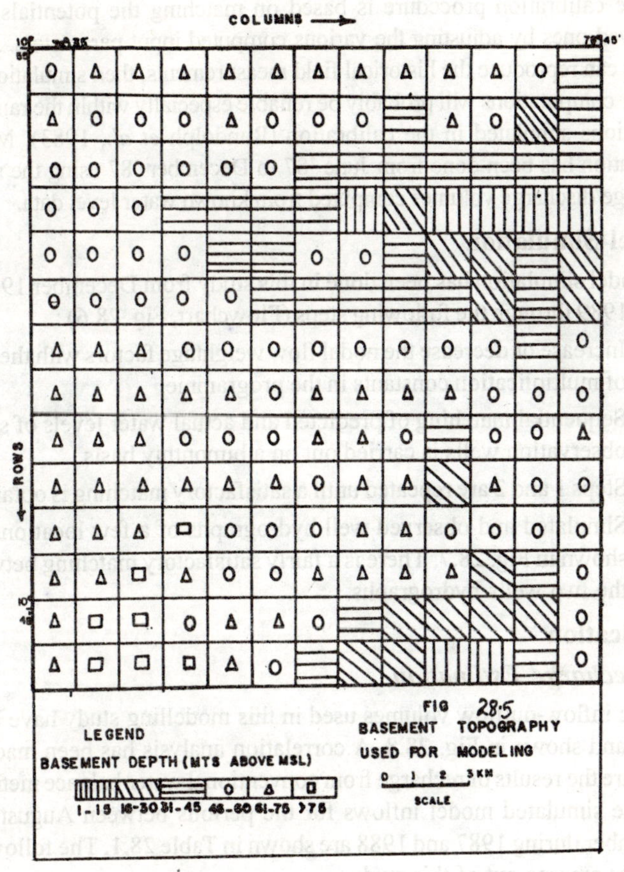

FIG 28·5
BASEMENT TOPOGRAPHY USED FOR MODELING

seasons. The values of this have been treated as flow weightage factors for the respective nodes. The time-step (Δt) used in all runs is 61 days.

Model Calibration

The calibration procedure is based on matching the potentials with observed ones by adjusting the various computed input parameters. If the model can reproduce the historical field measurements, then simulations of further computations will probably be reliable especially within the range of conditions simulated in the calibration (Randolph *et al.*, 1983). Model calibration has been done from June '87 to December '87 using the nodal recharge/discharge volumes computed from known water level data.

Model Simulation

Model simulation has been done in this study from December 1987 to April 1989 through the following steps (Flowchart, Fig. 28.6) :

1. Increase or decrease the nodal flow weightage factors with the help of multiplication constants in the programme.
2. Sequential matching of predicted and actual water levels of seven observation wells is carried out on a bimonthly basis.
3. Steps 1 and 2 are repeated until a satisfactory matching is obtained.

   Simulated and observed well hydrographs of a few locations are shown in Fig. 28.7. There is a fairly satisfactory matching between the two wells hydrographs.

Application

(1) Recharge Estimation

The inflow-outflow volumes used in this modelling study have been taken and shown in Fig. 28.8. A correlation analysis has been made to compare the results of recharge from conventional water balance methods and the simulated model inflows for the periods between August and December during 1987 and 1988 are shown in Table 28.1. The following features emerge out of this study.

(i) The inflows computed by the conventional method as well as model simulation are greater for the year 1988, a reflection on the increased rainfall contribution.

(ii) However, for both the years under consideration the computed model inflows are more than the anticipated rainfall recharge, probably due to the increment by baseflow.

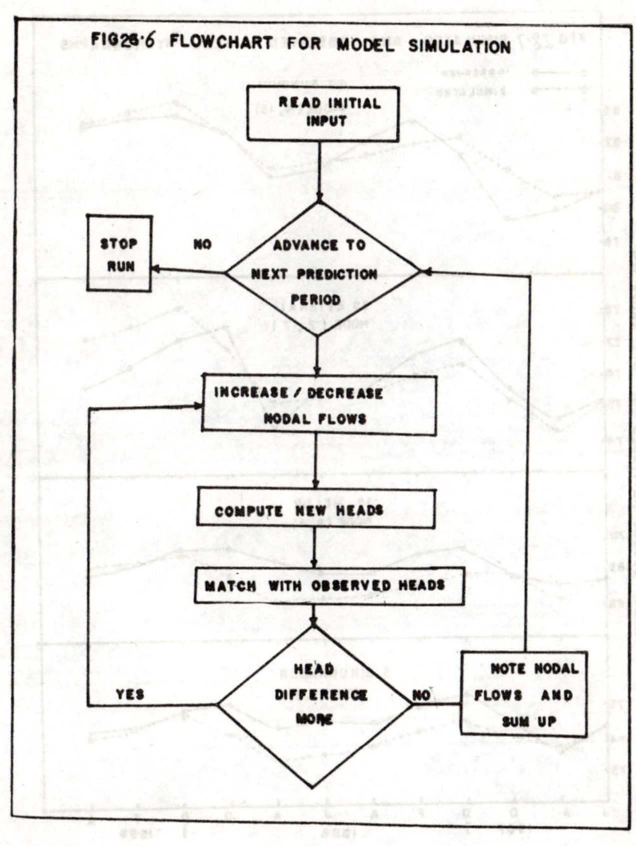
FIG 26·6 FLOWCHART FOR MODEL SIMULATION
READ INITIAL INPUT
STOP RUN
NO
ADVANCE TO NEXT PREDICTION PERIOD
INCREASE / DECREASE NODAL FLOWS
COMPUTE NEW HEADS
MATCH WITH OBSERVED HEADS
YES
HEAD DIFFERENCE MORE
NO
NOTE NODAL FLOWS AND SUM UP

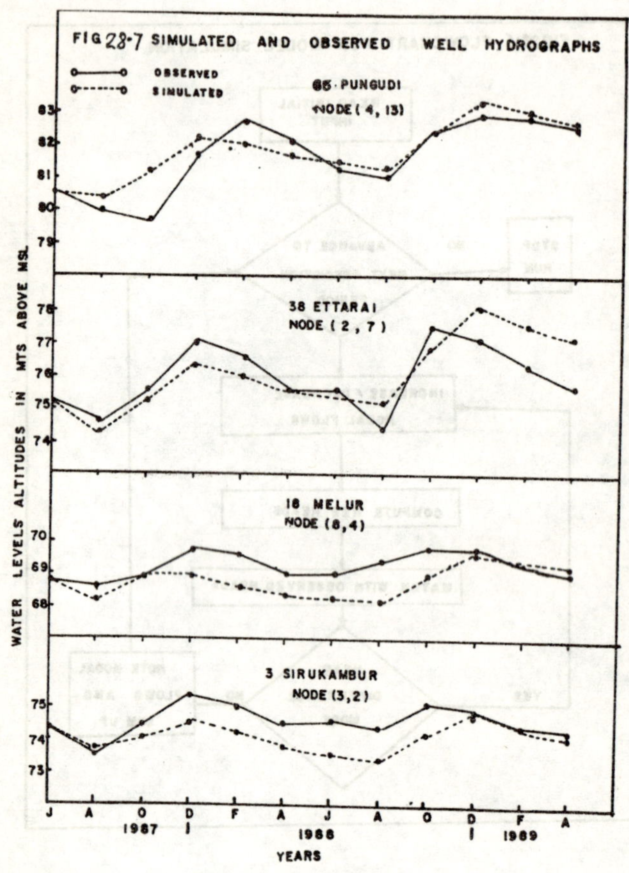
FIG 28·7 SIMULATED AND OBSERVED WELL HYDROGRAPHS
OBSERVED
SIMULATED
65 PUNGUDI
NODE (4, 13)
83
82
81
80
79
38 ETTARAI
NODE (2, 7)
78
77
76
75
74
16 MELUR
NODE (8,4)
70
69
68
3 SIRUKAMBUR
NODE (3,2)
75
74
73
WATER LEVELS ALTITUDES IN MTS ABOVE MSL
J A O D F A J A O D F A
1987 1988 1989
YEARS

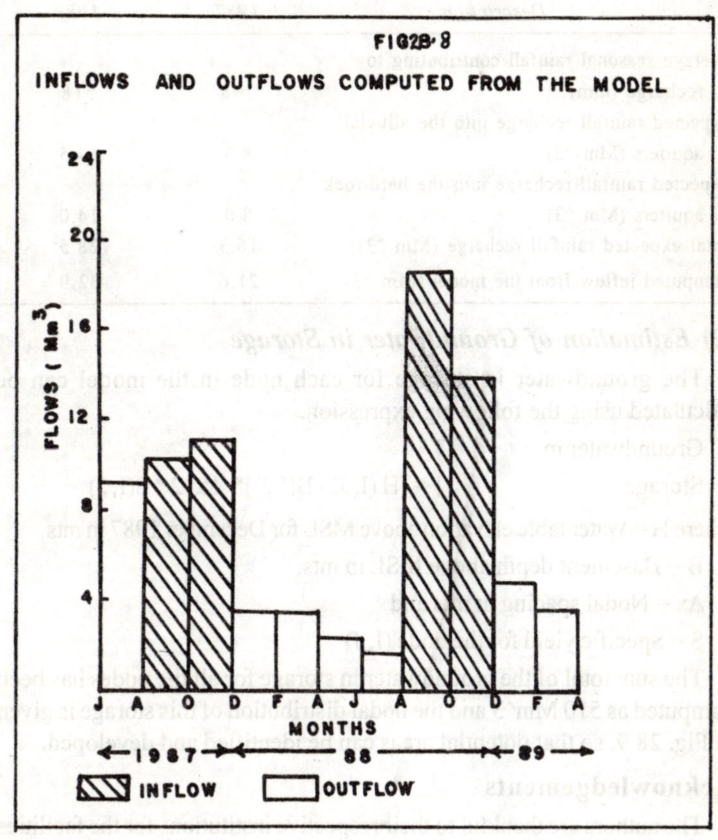
FIG28·8
INFLOWS AND OUTFLOWS COMPUTED FROM THE MODEL
FLOWS (Mm³)
24
20
16
12
8
4
J A O D F A J A O D F A
MONTHS
1987
88
89
INFLOW
OUTFLOW

Table 28.1 : Relation between Rainfall Recharge and Computed inflows

Description	*Aug.-Dec.* 1987	*Period* 1988
Average seasonal rainfall contributing to recharge (mm)	298	518
Expected rainfall recharge into the alluvial aquifers (Mm ^3)	8.3	14.5
Expected rainfall recharge into the hard-rock aquifers (Mm ^3)	8.0	14.0
Total expected rainfall recharge (Mm ^3)	16.3	28.5
Computed inflow from the model (Mm ^3)	21.6	32.9

(2) Estimation of Groundwater in Storage

The groundwater in storage for each node in the model can be calculated using the following expression.

$$\text{Groundwater in Storage} \Big\} = [H(I, J) - B(I\text{-}J)] * \Delta x\hat{}2 * S(I, J)$$

where H = Water table elevation above MSL for December 1987 in mts.

B = Basement depth above MSL in mts.

Δx = Nodal spacing in mts and

S = Specific yield for the node (I, J)

The sum total of the groundwater in storage for all the nodes has been computed as 510 Mm^3 and the nodal distribution of this storage is given in Fig. 28.9, so that potential areas can be identified and developed.

Acknowledgements

The authors are thankful to their respective institutions for the facilities provided during the course of this work. One of them (SM) would like to thank the UGC for granting the Teacher Fellowship under Faculty Improvement Programme.

References

Balasubramanian, A. and Sastri, J.C.V., (1989) Modeling the effects of proposed pumping from the Tambraparni River Basin, Tamil Nadu, India Inter. Nat. GW-89, NGRI, V. 2, pp. 629-40.

Prickett, T.A. and Lonnquist, C.G., (1971) Selected digital Computer techniques for groundwater resources evaluation, Illinois State Water Survey, Bull. 55.

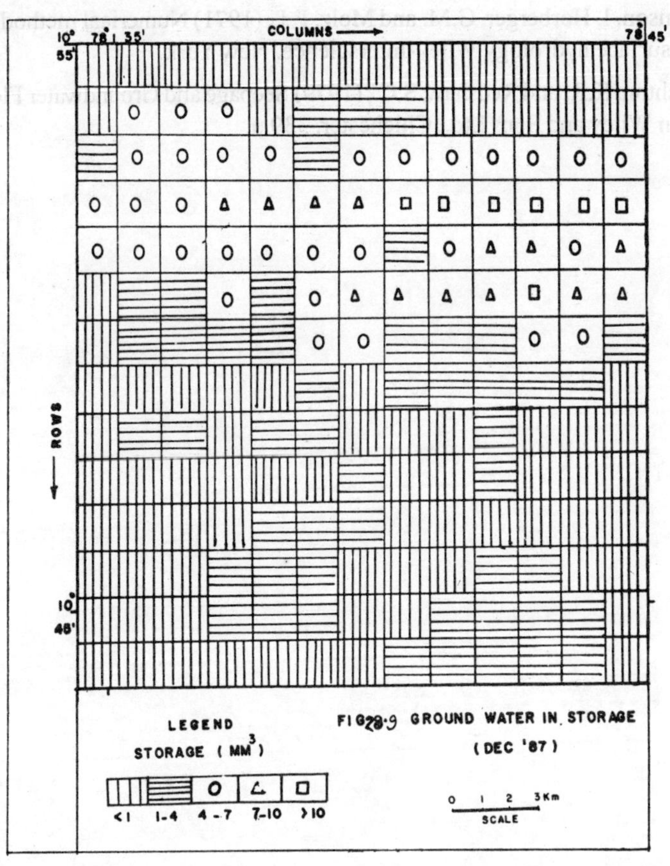

FIG 28.9 GROUND WATER IN STORAGE (DEC '87)

Randolph, R.B. and Krause, R.E., (1984) Analysis of the effects of proposed pumping from the principal artesian aquifer, Savannah, Georgia area. USGS Wat. Inv. Rept. 84-4064, 26 p.

Remson, I. Horberger, G.M. and Molz, F.J., (1971) Numerical methods in subsurface hydrology Wiley-Interscience, New York.

Rushton, K.R. and Redshaw, S.C., (1979) Seepage and Groundwater Flow, John Wiley and Sons Ltd., Chichester, 339 p.

Theory and Methods of Mathematical Modeling in Connection with the Evaluation of Hydrogeological Characteristics in Hard Rock Terrain of Bankura District, West Bengal

P.K. Das

Introduction

Khatra in Bankura district is situated at the south western part of West Bengal and is known to be drought prone area (Fig. 29.1). The area is underlain by hard crystalline metamorphic rocks of granite gneiss, amphibolite, quartzite of Precambrain age with undulating topography. The area is chiefly drained by Kasai river.

The objective of the present work is intended to study the evaluation of hydrologeological characteristics based on groundwater simulation.

Recently it found to be useful to apply the methods of mathematical modeling using suitable equation with some parameters for describing groundwater system. Also because of availability of micro computer the use of modeling has become widespread. Differential equations governing steady state or non steady state flow could not be applied earlier to real aquifer systems due to their complex character, geometry and boundary conditions. However, with the advancement of numerical methods and digital computers the solutions of these equations may readily be obtained to simulate the actual hydrologic system as well as to test the hypothesis. Further these have greatly expanded capabilities for solving predictive groundwater equation.

It may be considered from the historic well inventory data that an aquifer is in hydrogeologic balance, i.e., the amount of water moving into the system is approximately equal to the amount moving out of the system, although there may be some short term inflows and outflows which can affect the balance. Water may move into and out of the system in various ways such as stream flow, underflow, precipitation evapotranspiration, etc. Thus water level of an aquifer is a function of inflow-outflow differrence, and the hydrogeologic characteristics of the aquifer materails. The material characteristics which are related with water movement are

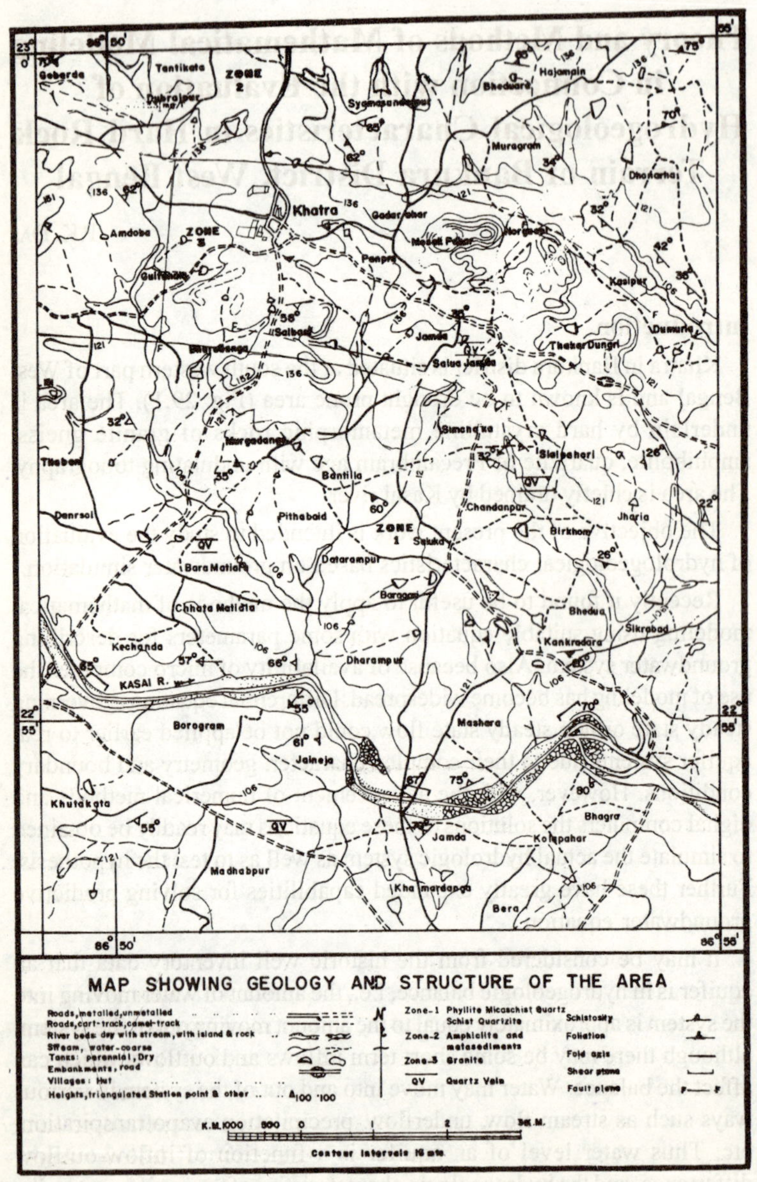
Geberda
Tantikata
ZONE 2
Sygmasandapur
Hajampin
Bhedwara
Muragram
Dhanarhar
Khatra
Amdoba
ZONE 3
Gutibhara
Panpra
Kasipur
Salbani
Jamda
Thakar Dungri
Dumuria
Simandura
Sialsahari
Thebari
Bantilla
Pithabaid
Chandanpur
Jharia
Danrsol
ZONE 1
Birkhom
Bara Matiala
Dataramput
Chhota Matiala
Bhedua
Sikrobad
Kechanda
Kankradara
Dharampur
KASAI R.
Borogram
Mashara
Jaineja
Khutakata
Bhagra
Jamda
Kolopahar
Madhabpur
Kha mardanga
Bera
MAP SHOWING GEOLOGY AND STRUCTURE OF THE AREA
Roads, metalled, unmetalled
Roads, cart-track, camel-track
River beds: dry with stream, with island & rock
Stream, water-course
Tanks: perennial, Dry
Embankments, road
Villages: Hut
Heights, triangulated Station point & other
Zone-1 Phyllite Micaschist Quartz Schist Quartzite
Zone-2 Amphicilite and metasediments
Zone-3 Granite
QV Quartz Vein
Schistosity
Foliation
Fold axis
Shear plane
Fault plane
K.M. 1000 500 0 1 2 3K.M.
Contour intervals 15 mts

Fig. 29.1

chiefly the porosity and permeability. The water flow in its natural state takes place from areas to higher to those of lower head in the aquifer.

The objective of the simulation is to determine the most probable hydrogeologic model when minimal field data are available. Based on simulation result the two most important aquifer properties, i.e., transmissivity (T) and storage coefficient/effective porosity (S) and water balance of the aquifer are presented.

Groundwater Flow Simulation

Flow Analysis

The differential equation which governs the non-steady homogeneous and isotropic flow in three dimension is given by Bittinger *et al.*, (1967).

$$\delta/\delta x\,(T\delta h/\delta x) + \delta/\delta x\,(T\delta h/\delta y) + \delta/\delta z\,(T\delta h/\delta z) = S\,\delta h/\delta t + Q \qquad (1)$$

Where T = aquifer transmissivity

h = head

t = time

S = aquifer storage coefficient

Q = net groundwater rate per unit area

X, Y, Z = rectangular coordinates.

As there is no solution in a closed form to the equation 1, a finite difference approach may be adopted for its numerical solution. By this approach the continuous aquifer system parameters are first replaced by an equivalent set of discrete elements. The equation expressing groundwater flow the discretized model is written in finite difference form. Finally the resulting set of finite difference equation is solved numerically with the aid of a digital computer.

A finite difference grid is superposed over a map of an aquifer which is subdivided into volumes, each having dimension m, x, y, m being its thickness. The grid line intersections are called nodes with a column (i) and row (j) in a coordinate system along x and y directions respectively. The continuity condition (Prickett and Lonnquist, 1968) relating the flow rates entering and leaving the nodes i, j (as shown in Fig. 29.2) requires that $Q_n + Q_1 + Q_3 = Q_2 + Q_4 + Q_6$ (2)

The portions of the aquifer included in the flow rate terms may be referred to as 'vector volumes' to emphasise that not only a volume but also direction of flow is being considered. Finally since time is discretized, equation 2 represents an instanteneous balance at the end of a time increment. Using Darcy's law the flow rate terms Q_1 through Q_4 can be written as,

Fig. 29.2

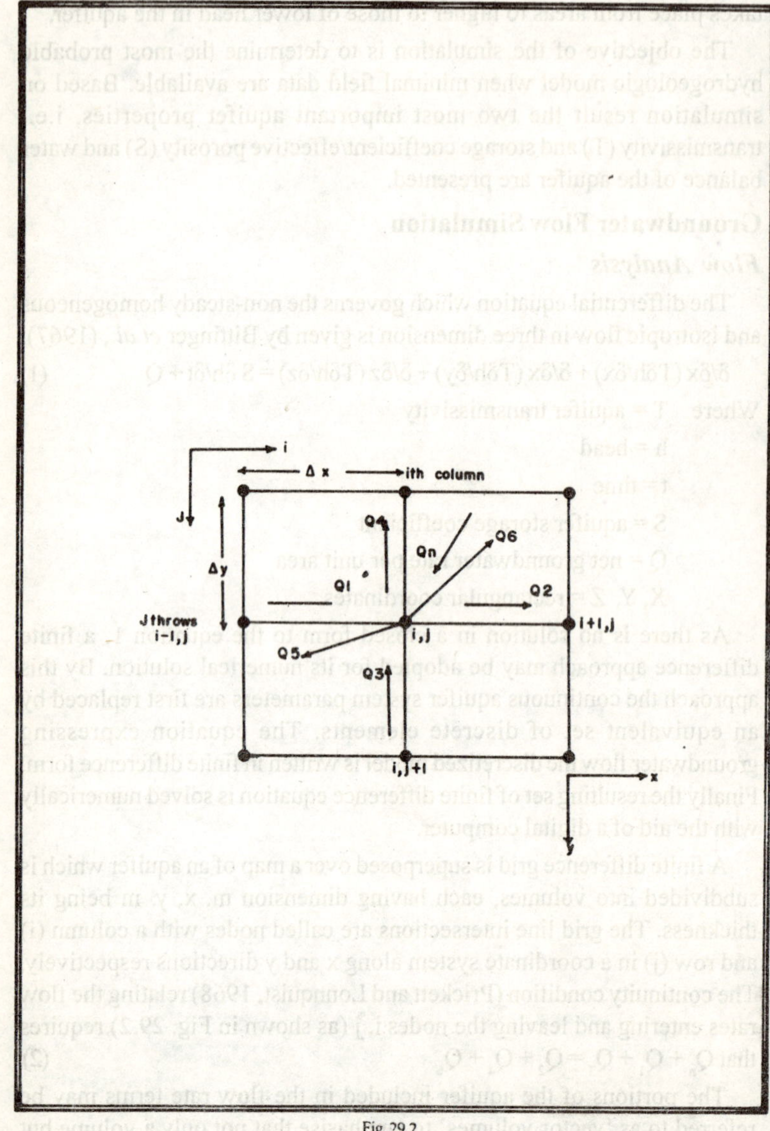

Fig. 29.2

$Q_1 = T_{i-1}, j_2 (h_{i-1}, j\text{-}h_j) \Delta y/\Delta x$ (3a)

$Q_2 = Ti, j, 2(h_i, j\text{-}h_{i+1}, j) \Delta y/\Delta x$ (3b)

$Q_3 = Ti, j (hi, j{+}1\text{-}hi, j) \Delta x/\Delta y$ (3c)

$Q_4 = Ti, j\text{-}1, 1 (hi, j\text{-}hi, j\text{-}i) \Delta x/\Delta y$ (3d)

Where $T_{i,j,1}$ = aquifer transmissivity within the vector volume between nodes i, j and is J + 1.

$T_{i,j,2}$ = aquifer transmissivity within vector volume between nodes i, j and i + 1, J.

$H_{i,j}$ = calculated heads at the ends of a time increment measured from an arbitrary reference level at nodes i, j.

The flow rate term Q representing the rate at which water is taken into storage is given by

$Q_5 = S\Delta * \Delta y (h_i, j\text{-}h\,\phi ij) \Delta t$ (4)

Where h_{ij} = Calculated heat at nodes i, j at the end of the previous time increment, t.

t = Time increment elapsed since last calculated heads.

The flow rate term Q_6 is made equal to a constant net withdrawal rate from vector volume of nodes i, j.

$Q_6 = Q_{i,j}$ (5)

The flow-rate term Q_n is set equal to zero when there is no special source or sink function.

Substitution of equations 3, 4, 5 & 6 into equation 2, then dividing both sides by Δx Δy and rearranging the terms result in

$T_{i\text{-}j}, j, 2(h_{i\text{-}j}, j\text{-}h_{ij}) / \Delta x^2 + T_i, j, 2(h_{i+1}, j\text{-}h_i) \Delta x^2 + T_i, j, 1 (h_i, j{+}1\text{-}h_i, j) / \Delta y^2 +$
$T_i, j\text{-}1, 1 (h_i, j\text{-}1\text{-}h_i, j) \Delta y^2$
$= S (hi, j\text{-}h\phi i, j) \Delta t + Q_t, J/\Delta x.\Delta y\text{-}Q_n/\Delta x.\Delta y$ (6)

The equation 6 representing the finite difference of the equation 1 can thus be solved by a modified form of the Alternating Direction Implicit Procedure (ADIP) of Peaceman and Rachford (1955). The node equations are solved by Gauses elimination of individual columns one by one while those for rows are held constant. The interaction process is then completed after all equations are solved row by row and is repeated number of times to achieve convergence thus giving the calculation for the given time increment.

Model Design and Data Input

The area has been divided into cells defined by 19 rows (X) and 22 columns (Y), each cell being 500 m in X direction and 500 m in Y direction.

The total model area is stretched over an area of 104.5 sq. km. comprising 418 cells. It is important to specify the boundary conditions in conceptualizing groundwater flow system. The river Kasai and the two creeks, Kanria Jhor and Mura Jhor form natural outlets for groundwater flow to the south, to the west and to the east of the model area respectively. Their dispositions are shown in Fig. 29.3a. There is no natural boundary to the north of the areas, Row 1 (Fig. 29.3a) is considered to represent the northern boundary of the study area and the model but not the physical end of groundwater system and as such it has a specific role being the artificial boundary of the model. It is simulated as a constant head boundary which takes care of subsurface inflow into the model area from cut-off part of the area. The steady water table as obtained from the dugwell inventory of the post-monsoon and pre-monsoon periods may be considered to represent the upper free surface and lower boundary surface of the model. The aquifer simulation program in Fortran IV is used to solve equation 6. The programme includes also subroutines for printing map of water level and calculation of water balance. The hydrologic parameters which describe the water table condition of the area comprise: (*i*) transmissivity and (*ii*) storage coefficient/effective porosity. It may be assumed that the other major factors controlling the heads are : (*i*) recharge from precipitation and (*ii*) artificial discharge through pumping.

One of the primary objectives of groundwater investigations is to determine the hydrogeological parameters for example, transmissivity and effective porosity. The traditional method to determine these parameters is pump tests but these tests are difficult and very expensive to conduct. Fortunately inverse method of parameter identification are available which begin with the preparation of groundwater model that incorporate the best available values of T and S. Thus sets of T and S values are input into computer which is also provided with sets of probable values of recharge and discharge values. Using these values the model thus generates groundwater levels which are compared with observed water levels. Adjustments are made in these input values until best agreement is achieved between the generated and observed head values. The solution may not be unique but be left to the modeler who must be experienced hydrogeologist to evaluate which set of data is most realistic for a particular area.

Evaluation of Groundwater System Based on Simulation

Considering the area, which is characterised by hard rocks several sets of suitable and reasonable values of transmissivity, storage

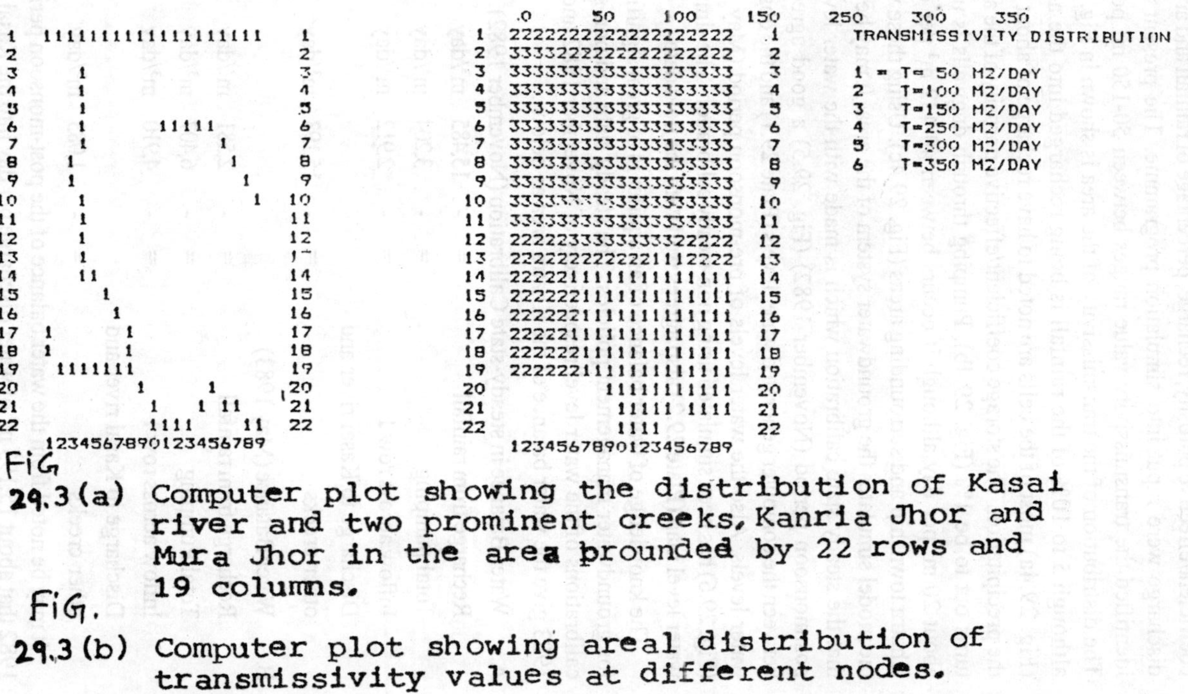

FIG. 29.3 (a) Computer plot showing the distribution of Kasai river and two prominent creeks, Kanria Jhor and Mura Jhor in the area prounded by 22 rows and 19 columns.

FIG. 29.3 (b) Computer plot showing areal distribution of transmissivity values at different nodes.

coefficient/effective porosity, recharge, percentage of rainfall and artificial discharge were input into simulation programme. The present model identified the transmissivity value ranges between 50-150 m^2 per day. The distribution of the transmissivity of the area is shown in Fig. 29.3b although 5 to 10% of the rainfall is being recharged into the aquifer (Fig. 29.4a) most of the cells are noted to have recharge of about 5% of the precipitation. The storage coefficient/effective porosity of the aquifer turns out to be 1% (Fig. 29.4b). Pumping through dugwells is usually about 20 m^3 per day although it occurs between 40-432 m^3 per day at Khatra township and surrounding areas (Fig. 29.4c). Using these values the model simulates the groundwater system of the area. It may be noted that the steady state calibration which is made with the water levels of postmonsoon period (November 1982) (Fig. 29.5) a good agreement between the computer generated water levels (Plate 29.1) and the observed water levels. Also the water levels of pre-monsoon period (May 1983) (Fig. 29.6) has been simulated using non-steady calibration. The simulated water level map (Plate 29.2) also agrees well with the field data.

The knowledge of water balance of an aquifer is of great significance for groundwater management policies. The steady state and non-steady calibrations of the water level respectively of November 1982 and May 1983 give the water balance of the Khatra area as below :

A. Water Balance in steady-state Calibration (November 1982).

Recharge from rainfall	= -	15,485 m^3/day
Total pumping	= -	3,204 m^3/day
Inflow across row 1	= -	2,912 m^3/day
Discharge to Kasai river and other creeks	= -	15,193 m^3/day

B. Water Balance (May 1983)

Recharge from rainfall	= -	2,581 m^3/day
Total Pumping	= -	6,407 m^3/day
Inflow across row 1	= -	4,910 m^3/day
Discharge to Kasai river and other creeks	= -	1,083 m^3/day

It may be noted from the water balance of the post-monsoon period of 1982 that about 15,193 m^3 of water discharge per day into the Kasai river. This quantity of water could of course be pumped out of the aquifer but because of monsoon time additional water may not be required except for

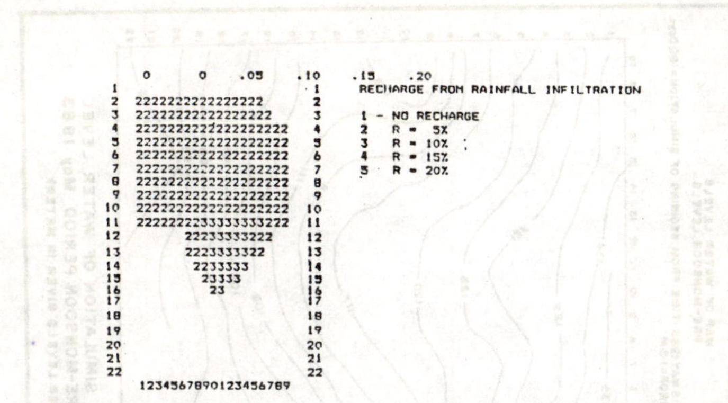

Fig. 29.4 (a) Computer plot showing the areal distribution of recharge from rainfall.

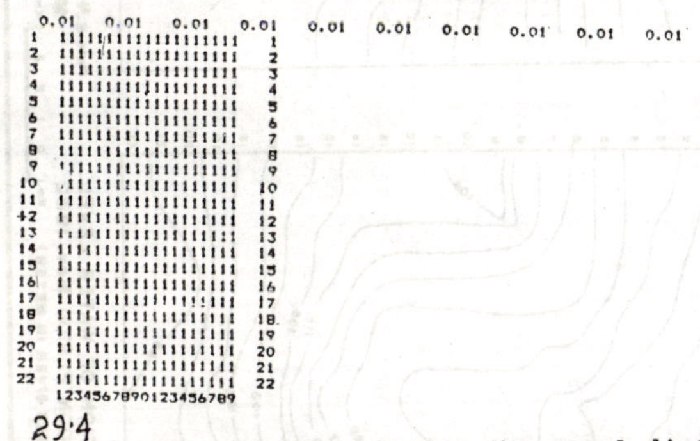

Fig. 29.4 (b) Computer plot showing the areal distribution of storage coefficient effective porosity.

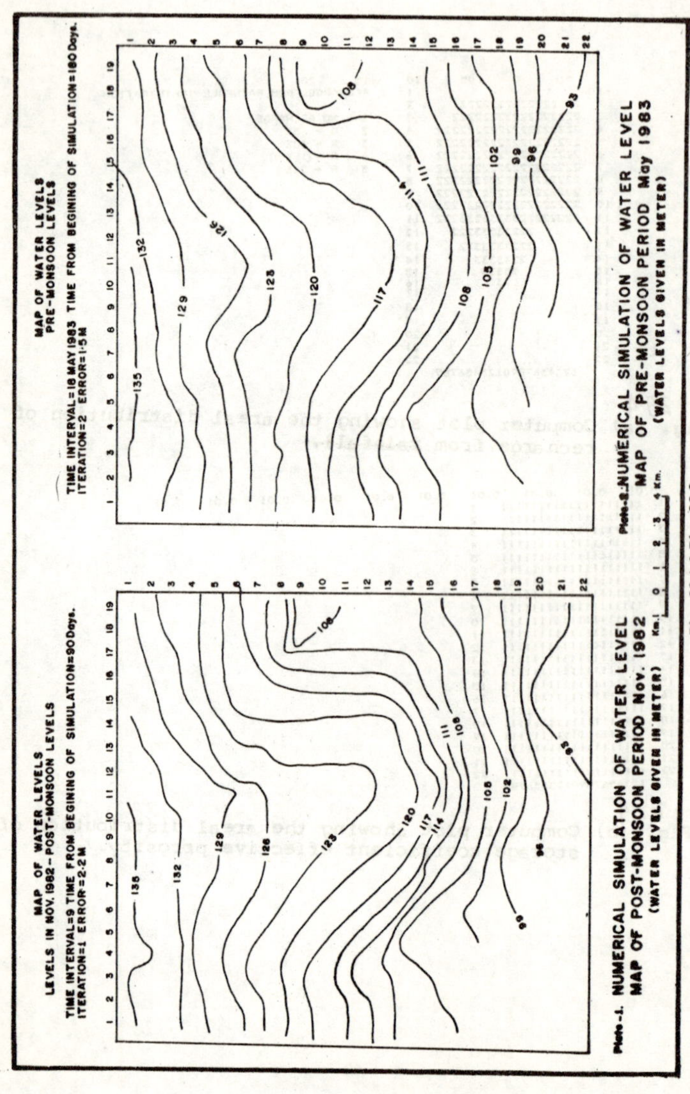

Plate-1. NUMERICAL SIMULATION OF WATER LEVEL MAP OF POST-MONSOON PERIOD Nov. 1982 (WATER LEVELS GIVEN IN METER)

Plate-2: NUMERICAL SIMULATION OF WATER LEVEL MAP OF PRE-MONSOON PERIOD May 1983 (WATER LEVELS GIVEN IN METER)

Plate 29.1 & Plate 29.2

```
         0    .24    .48    .72     1.0    2.0     5     7     8     10
   1                           1
   2  1  22  2         11      2     PUMPING FROM DUG WELLS
   3      22 2 . 1 1 1 1       3
   4           2               4     1 - Q=0.24 L/S  (20.0 M3/DAY)
   5  1     4 4  11            5     2 - Q=0.48 L/S  (40.0 M3/DAY)
   6 11  246  4                6     3 - Q=0.72 L/S  (60.0 M3/DAY)
   7  1 123 3 21  1            7     4 - Q=1.00 L/S  (86.4 M3/DAY)
   8  2    121 1  1            8     5 - Q=2.00 L/S  (173  M3/DAY)
   9    11   1        11       9     6 - Q=5.00 L/S  (432  M3/DAY)
  10    1 2 1  111      1     10     7 - Q=7
  11   211 1    1             11     8 - Q=8
  12   1  1  1 1  1 1         12     9 - Q=10
  13   1   1 1  11  1 1       13
  14   1  11  2               14
  15   1  1  1  1 1 1 2       15
  16   1  11   1  1           16
  17      21    11  1   1     17
  18     1  1       1   11    18
  19           2        1     19
  20           1 1            20
  21                          21
 '22                          22
      1234567890123456789
```

Fig. 29·4C Computer plot showing areal distribution of Pumping from Dug Wells.

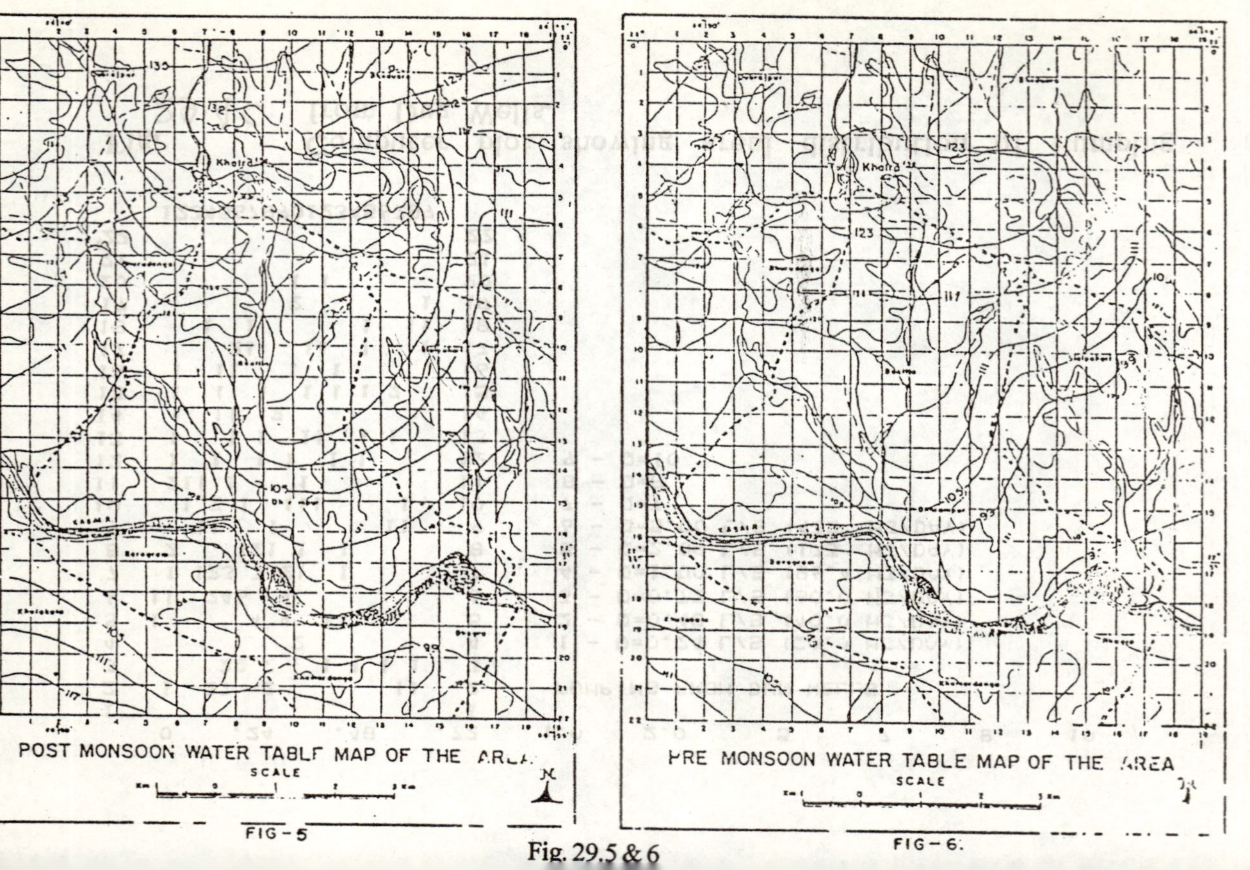
POST MONSOON WATER TABLE MAP OF THE AREA
SCALE
FIG-5
PRE MONSOON WATER TABLE MAP OF THE AREA
SCALE
FIG-6.

Fig. 29.5 & 6

special purposes, for example irrigation and industry. Gradual decrease in discharge takes place after the monsoon period which amounts to 1,083 m^3 per day. This quantum of water is, however, no less significant considering the water crisis in the area even for drinking purpose. This discharge can, however, be prevented by sinking wells which should be strategically sited by the side of the river and the creeks. Thus, water supply can be enhanced through 50 dugwells each producing at least 20 m^3/day.

For predictive purpose the water level map of the post-monsoon period (November 1982) (Plate 29.3) has been constructed by the model using the hydrogeological data is identified by the computer. The water balance for November 1982 is given below.

Water Balance in Steady-State Calibration (November (1982))

ISTEP = 1	Recharge from rainfall	=	- 15,485	m^3/day
	Total pumping	=	3,204	m^3/day
	Inflow across row 1	=	- 2,912	m^3/day
	(and other creeks)	=	15,193	m^3/day

Size of the model : 19 columns, 22 rows spacing in X and Y directions : 500 m

Size of one cell : 500 * 500 m

Total area of the model : 104.5 sq. km.

Water Balance

ISTEP = 18	(May 1983)			
	Recharge from rainfall	=	- 2581	m^3/day
	Total pumping	=	- 6407	m^3/day
	Inflow across row 1	=	- 4910	m^3/day
	Discharge to river Kasai			
	(and other creeks)	=	- 1083	m^3/day
ISTEP = 36			(November 1983)	
	Recharge from rainfall	=	- 13420	m^3/day
	Total pumping	=	- 3204	m^3/day
	Inflow across row 1	=	- 2719	m^3/day
	Discharge to river Kasai			
	(and other creeks)	=	- 12935	m^3/day

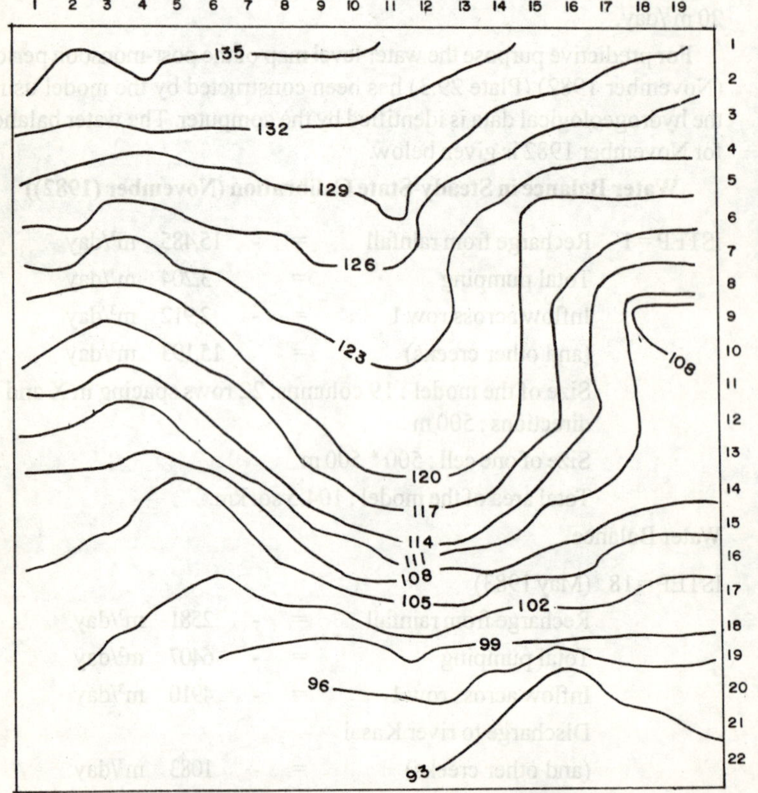
MAP OF WATER LEVELS
POST-MONSOON LEVELS
TIME INTERVAL=36 Nov.1983 TIME FROM BEGINING OF SIMULATION= 360 Days.
1 2 3 4 5 6 7 8 9 10 11 12 13 14 15 16 17 18 19
1 2 3 4 5 6 7 8 9 10 11 12 13 14 15 16 17 18 19 20 21 22
135
132
129
126
123
108
120
117
114
111
108
105
102
99
96
93
NUMERICAL SIMULATION OF PREDICTIVE MAP
OF POST-MONSOON PERIOD NOVEMBER 1983
(WATER LEVELS GIVEN IN METER)
km. 1 0 1 2 3 4 km.

Plate 29.3

Results and Discussion

The digital model which has been constructed to simulate the observed hydraulic heads of the area gives a quantitative appraisal of the unconfined aquifer system. The model, however, can be no more accurate than the information upon which it is based. The hydraulic head has been calculated from the well inventory data and theodolite survey. As the hydrogeologic parameters for example T and S are difficult to determine especially in this type of rock terrain a wide range of values as may be appropriate for hard rock terrain have been input into the computer. The model gives the area distribution of these two important parameters as well as the recharge by rainfall and the discharge through pumping. This model also yields the water balance of the area which will facilitate the management of the water of unconfined aquifer. The water balance of the area gives an estimate of some water which is being discharged into the Kasai river even during summer time. This discharge can be prevented by suitably placing wells by the side of the Kasai river and the two creeks, i.e., Mura Jhor and Kanria Jhor, thus indicating the usefulness of the model for water management.

Acknowledgements

The author is gratefully indebted to Prof. S. Basu Mallik, (Head of the Department, Geological Sciences, Jadavpur University, Calcutta) under whose guidance this paper has been prepared. The author gratefully acknowledges the support of Science and Technology Committee, Govt. of West Bengal for the award of research fellowship.

References

Bittinger, M.W., Duke, H.R. and Logenbaugh, R.A., (1967) Mathematical simulation for better aquifer management Intern. Assoc. of Scientific Hydrology, Symposium of Haifa Publication 72.

Peaceman, D.W. and Rachford, H.H. Jr., (1955) The numerical solution of parabolia and elliptic differential equation. Jour. Soc., of Industrial & Applied Mathematics, V. 3, p. 28.

Prickett, T.A. and Lonnquist, C.G., (1968) Aquifer simulation Programme listing using Alternating Direction Implicit Method. Illinois State water Survey mimieographed report presented at Int. Assoc. of Scientific Hydrology Symposium on use of computers in Hydrology, Tucson, Arizona.

30

Watch – A Fortran 77 Program for Processing the Analytical Hydrogeochemical Data

A. RAMANATHAN
AND
D. CHANDRASEKHARAM

Introduction

There are programs like WATQF available for calculating the chemical equilibrium and other chemical thermodynamical functions. HYDROCHEM is another FORTRAN IV program (Gill and Rusenthal, 1975), which can process the hydrogeochemical data. There is a need in the air for a simple programme which could check and process the analytical geochemical data. So an attempt is made to write this program called WATCH (water chemistry). The description of the program and other details regarding, working with the program is explained in the following paragraph.

Chemistry of Water

The common constituents of water are :

Alkalies : Na, K, Li

Alkaline earths : Ca, Mg, Sr, Ba

Halide : Cl

Carbonates : CO_3, HCO_3 and

Sulphate : SO_4

Water quality is determined by analytically measuring the concentration of these various constituents and the effect or properties caused by those substances.

Program Description

Program WATCH is written in FORTRAN 77 for the Unix/5.2 computer of Department of Earth Sciences, I.I.T., Bombay and it requires 17.5 K of storage. It takes only 3 seconds of execution time. The program is written in a simple language with a very few of complicated structures like go to statements.

Program WATCH is divided into a series of subroutines for clarity and efficiency, *main* reads the basic data and manages the other subroutines;

unit changes the unit of given concentration of elements to other units; *dcheck* the data for analytical error and percentage of error of present; *prop* calculates all the properties of water (collected from various literature) regarding water quality; *irat* provides the ionic ratio of desirable elements; *ctherm* is for calculating chemical thermo-dynamical aspects from the given set of data.

The general structure of the whole program and the way it works is shown in Fig. 30.1.

(i) Basic Data Used

The basic data used are the name of the species, its charge, gram formula weight and Debye-Huckel parameter. The important major ions and some minor ions which are expected commonly in the water are taken for study. The Debye-Huckel parameter is taken from Robie *et al*., (1978). The data is shown in Table 30.1.

(ii) Input

The input to the program is the name of the sample, temperature, field p^H, lab p^H, conductivity and analytical concentrations of ions as per the order of basic data. Concentration of the ions can be given in any of the following units : (1) ppm, (2) epm and (3) molality.

(iii) Optional Input

The number of samples wanted to be run can be given in the data. The wish of calling the different subroutines, which are described earlier is given to the user. 'Cflag' is a flag used for the purpose. If one puts of the 'cflag' then only the given data will be printed and no other functions will be performed. Likewise, there is another variable, 'prflag', which acts like an 'on-off' switch for optional printing.

(iv) Program

The program with the different subroutines and its functions are explained below :

(a) Main : Main program reads the basic data and after checking the user's option it calls the different subroutines. It prints the basic data if the user wish to have it.

(b) Unit : This subroutine converts the given unit of concentration to the other two units and stores it. Using the prflag variable, already explained, the concentration of all the ions in the given data, can be obtained on all the three units.

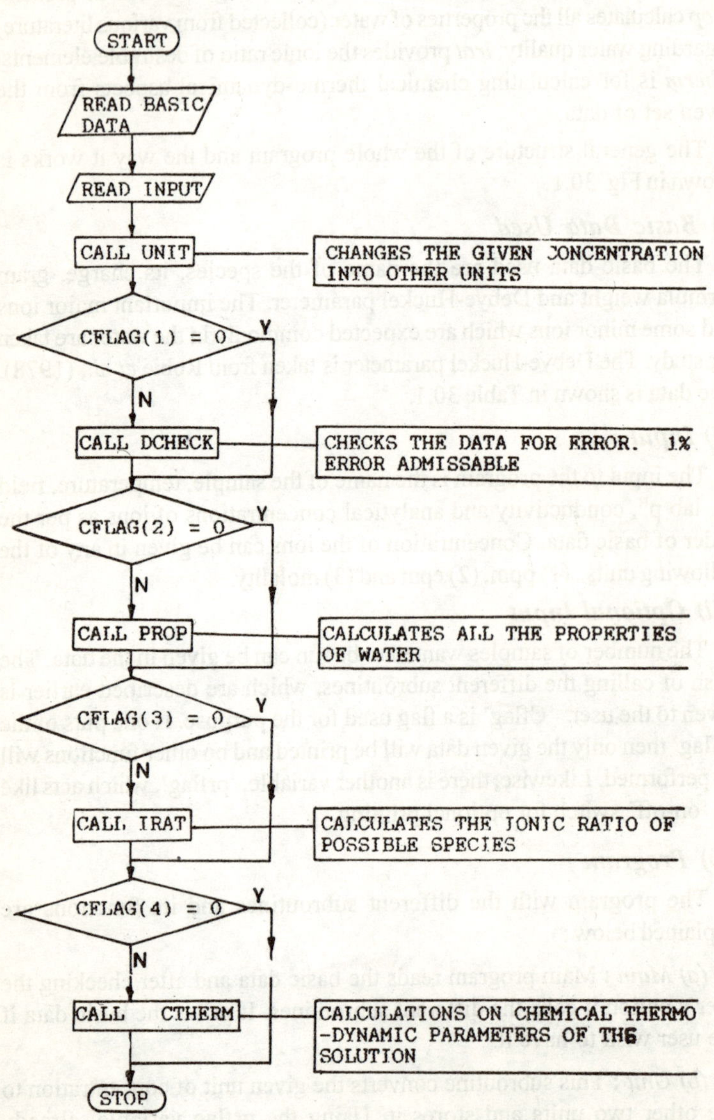

FIG.30·1 BASIC STRUCTURE OF THE PROGRAM

Table 30.1 : Basic Data

'CA	' 2,	40.080,6.0
'MG	' 2,	24:3120,6.5
'NA	' 1,	22.9898,4.0
'K	' 1,	39.1020,3.0
'CL	' -1,	35.4530,3.0
'SO_4	' -2,	96.0616,4.0
'HCO_3	' -1,	61.0173,5.4
'SIO_2TOT	' 0,	60.0848,0.0
'FE	' 2,	55.8470,6.0
'FE	' 3,	55.8470,9.0
'CO_3	' -2,	60.0094,5.4
'F	' -1,	18.9984,3.5
'H	' 1,	1.0080,9.0
'L!	' 1,	6.9390,6.0
'MN	' 2,	54.9400,6.0
'NO_3	' -1,	62.0049,3.0
'OH	' -1,	17.0074,3.5
'BR	' -1,	79.9090,4.0

(c) dcheck : This subroutine calculates the total of all anions and cations (epm unit) and calculates the difference. If the difference is above 1% then it prints a error message and returns back to the main program. The percentage of error is also calculated as given below :

% of error = (difference between cation and anion/cation + anion) * 100

(d) prop : This subroutine calculates the properties from the given data. The water is classified based on the amount of TDS present (Todd, 1980). The temporary hardness, permanent hardness and the total hardness are calculated using the standard formulae. The quality factors of water for irrigation such as potential soil salinity (pss), permeability index (perind) (Garg, 1982), Na%, SAR and RSC (Awasthi, 1974) are calculated and classified whether water is good or bad for irrigation. The TDS, conductivity and the concentration of Cl are also studied to classify the water for irrigation. The respective formulae are given below. The limits to classify the water can be referred from the cited literature.

pss = Cl + (SO_4/2)

perind = Na + sqrt (HCO_3) Ca + Mg + Na * 100

% Na = (Na + k) * 100 / (Ca + Mg + Na + K)

RSC = (HCO_3 + CO_3) - (Ca + Mg)

(Concentrations in epm unit)

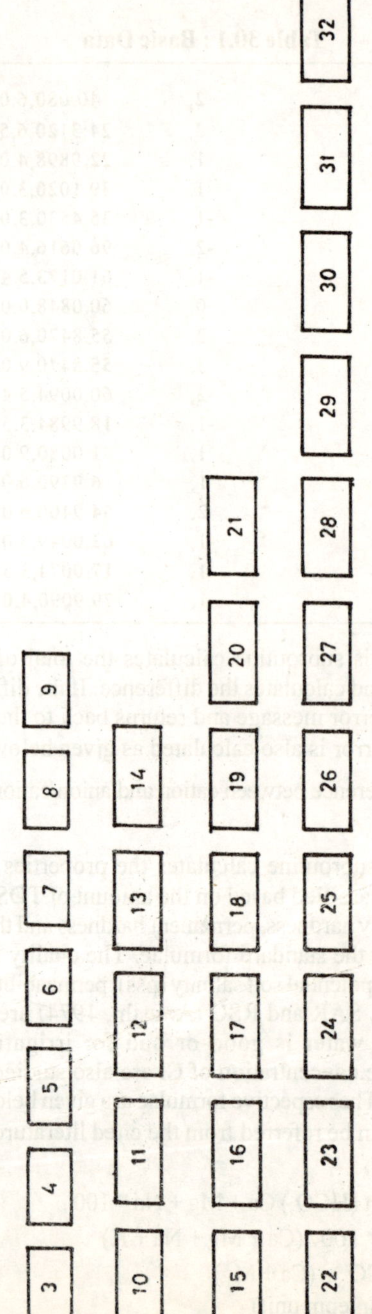

FIG 30.2: INPUT AND OPTIONAL INPUT

Valoyskha factor used by Soviet Scientist's is also calculated. The salinity of the water can be interpreted form this factor. The ion exchange calculations are done for both positive and negative ions. Thus this subroutine produces valuable and publishable results.

(e) irat : It gives the indices of ionic ratio of some of the elements. Hence a variable interpretation is possible, and it is left to the user for interpretation. So this subroutine prints just the different ratios calculated in epm unit. However, a brief note is added here to explain some of the obvious uses of the some of the ratios. HCO_3-Cl ratio can be used to draw contours around the sample points to find the recharge area. Br-Cl ratio is used to find out whether the water encountered any evaporties on its travel. Mg-Ca ratio is very useful in finding out the type of carbonate which is in saturation with the water which in turn indicates the type of terrain.

(f) Ctherm : This subroutine calculates the ionic strength of the water from the given set of analyses. Ionic strength is one step ahead of concentration because the term includes the charge which is an electrical property. It calculates the activity coefficient, very important thermodynamic factor for all the elements analysed. This is performed using Debye-Huckel's equation, which is given below.

$$-\log r = A.\ z^2.\ \text{squrt}\ (i) / (1 + B.\ \text{squrt}\ (i))$$

where, z is the charge of species, A and B are constants whose value changes with temperature and i is the ionic strength. This equation holds good for solutions of ionic strength less than 0.1. However, the same equation holds good to an approximations for solutions whose ionic strength is more than 0.1. There is a parallel equation dealt by Davies. Here, Debye-Huckel's equation is used. The constants A and B are available from Garrels *et al.*, (1965). This provides data for a difference of temperature of 5°, from 0 degrees to 60 degrees. Here, an interpolation subroutine (intpn) is used to find out the constants A and B for the temperature which comes in between. Since the data starts from 0 degrees and there is no negative temperature possible the subroutine intpn is made for calculating for temperatures above 0 degrees. Temperature above 60 degrees will return to main function as it did in the case of dcheck and terminates program thereof. Thus the constants obtained are used in the above equation to find out the activity coefficient of the ions in equation.

The solubility of calcium carbonate is determined in large part by the pH of its environment and that of gypsum depends on the terrain the water travels or rests and the time of stay. The equilibrium constant (K) values change with temperature and by using recent data this value is calculated for the given temperature. Saturation index for calcite and gypsum is also made available adopting the methodology of Langumir (1971). It is explained briefly here.

(1) Ionic strength of the solution and molality of the ion and the activity coefficient of that ion in question from the previous steps are used here.

(2) Calculate the actual ion product (iapc and iapg) in the sample solution.

(3) Calculate the equilibrium constant for the given temperature. Here, the expression given by Plummer and Busenberg (1982) is used.

(4) Find out the ratio of the actual ionic product that is the original K to the theoretical K, the real value expected at equilibrium. If both are same, the value of the ratio will be 1 and the log of it will be zero.

(5) Assumption of certain values make + or -0.1 error, so a value between -0.1 to + 0.1 would indicate that the given pair of ions is in saturation. A value more than 0.1 and a value less than -0.1 would indicate oversaturation and undersaturation respectively. Using this methodology, the saturation index (sic and sig) for calcite and gypsum are calculated. This is shown in the model run of the program.

Parallely, there are other methods (Scholelar, 1956) but the authors prefer to use this methodology, with the set of new data, because of its simplicity. Similarly, there are number of methods like Vanthoff's equation for calculating the value of K for different temperatures, but for the same reasons, Plummer's expressions are used in this program.

Output

By properly using the flags one gets : (i) only the given data as output and/or (ii) one or two or all the subroutines, basic data used in the program and thereby results, can or can't be printed.

Flexibility of the Program

Program written for zenith is easily convertible to p.c. mode. The data used can be replaced with advent of new data. Some more elements of user's interest can be added to the basic data. Some more calculations or functions can be added to every subroutine. Since the program is in FORTRAN 77 language and written with less complicated structure, it is easily readable and highly flexible. Several data software run on this program and found successful. But dissolved gases in water are not considered in the program. The physical properties are also not given importance. These two can be rectified by improving the program.

An Example

For clarity, a model data has taken for processing and the results are obtained with explanations (Appendix 1).

Acknowledgements

Authors thank the Director of the Institute and Head of the Department of Earth Sciences for providing facilities for doing this work.

References

Awasthi, S.C., (1974) Quality of water in relation to its use. Indian Minerals. V. 27, No. 3, pp. 27-38.

Garg, S.P., (1982) Groundwater and tube wells. 2nd edition Oxford & IBH Publishers Company, New Delhi.

Gill, D. and Rusenhal, E., (1975) HYDROCHEM - A FORTRAN-IV Program for processing analytical hydrogeochemical data. Computers and Geosciences, V. 1, pp. 83-96.

Langumir D., (1971) The geochemistry of some carbonate groundwaters in central Pennsylvania. Geochem. Cosmochem. Acta. V. 35, pp. 1023-45.

Plummer, L.N. and Busenberg, E., (1982) The solubilities of calcite, aragonite and vaterite in CO_2-H_2O solutions between 0 to 90°C and an evaluation of the aqueous model for the system $CaCO_2$-CO_2-H_2O. Geochim. Cosmochem. Acta. V. 46, pp. 1011-40.

Robie, R.A., Hemingway, B.S. and Fisher, J.R., (1978) Thermodynamic properties of minerals and related substances at 298k and 1 bar pressure and at higher temperatures. U.S. Goel. Surv. Bull. P. 1452, 456 p.

Scholelar, H., (1956) Geochimie des eaux souterraines : Technip, Paris, 213 p. Todd, D.K., (1980) Groundwater hydrology. John Wiley & Sons, New York, 535 p.

Appendix - I

To illustrate the execution, a model data was run on this program. The details are given below.

Input data (given below) includes basic data and the analytical data of math hot spring, the same data has been run on 2 different flag conditions, one on, all the flags open and another on all the flags closed.

'CA	'	2,	40.080,6.0
'MG	'	2,	24,3120,6.5
'NA	'	1,	22.9898,4.0
'K	'	1,	39.1020,3.0
'CL	'	-1,	35.4530,3.0
'SO_4	'	-2,	96.0616,4.0
'HCO_3	'	-1,	61.0173, 5.4
'SiO_2TOT	'	0,	60.0848,0.0
'Fe	'	2,	55.8470,6.0
'Fe	'	3,	55.8470,6.0
'CO_3	'	-2,	60.0094,5.4
'F	'	-1,	18.9984,3.5
'H	'	1,	1.0080,9.0
'LI	'	1,	6.9390,6.0
'MN	'	2,	54.9400,6.0
'NO_3	'	-1,	62.0049,3.0
'OH	'	-1,	17.0074,3.5
'BR	'	-1,	79.9090,4.0

'Math hostp'

0, 1, 1, 1, 1, 1, 1

50.0, 8.4, 8.45, 509.0, 1100.0

4.0, 0.001, 125.0, 6.0, 69.0, 57.0, 154.0

130.0, 0.0001, 0.0001, 3.0, 0.0001, 0.0001, 0.0001, 0.0001, 0.0001, 0.001

'Math hostp'

0, 0, 0, 0, 0, 0, 0

50.0, 8.4, 8.45, 509.0, 1100.0

4.0, 0.001, 125.0, 69.0, 57.0, 154.0

130.0, 0.0001, 0.001, 0.0001, 3.0, 0.0001, 0.0001, 0.0001, 0.0001, 0.0001

Results of the Data Processed on all Flags Open

Logical Inferences from Hydrogeochemical Data

Number of datas wanted to be processed = 2

Name of the sample = Math hostp

Unit flag = 0 print flag = 11 call flag = 1111

Physical Parameters

Temperature = 50.00

Field p^H = 8.40

LAB p^H = 8.45

Total dissolved solids = 509.00

Conductivity = 1100.00

Data of elements and their concentration

'CA	'	4.000
'MG	'	.001
'NA	'	125.000
'K	'	6.000
'CL	'	69.000
'SO_4	'	57.000
'HCO_3	'	130.00
'SIO_2TOT	'	.000
'FE	'	.000
'FE	'	.000
'CO_3	'	.000
'F	'	3.000
'H	'	.000
'LI	'	.000
'MN	'	.000
'NO_3	'	.000
'OH	'	.000
'BR	'	.000

The Given Data in the nits PPM, EPM and Molality

ppm	*epm*	*molality*
4.00000	.19960	.00010
.00100	.00008	.00000
125.00000	5.43719	.00544
6,00000	.15344	.00015

69.00000	1.94624	.00195
57.00000	1.18674	.00059
154.00000	2.52387	.00252
130.00000	.00000	.00216
.00010	.00000	.00000
.00010	.00001	.00000
.00010	.00000	.00000
3.00000	.15791	.00016
.00010	.00010	.00000
.00010	.00001	.00000
.00010	.00000	.00000
.00010	.00000	.00000
.00010	.00001	.00000
.00010	.00000	.00000

Total of anions = 5.8

Total of cations = 5.79

The difference between the cations and the anions = -.02

% of error = -.21

Error below the limit

Salinity Factor	**Suitability**
Total dissolved solids = 509.00	fresh water

Hardness	**Class**
Total hardness =	.50
Temporary hardness =	126.31
Permanent hardness =	-125.81

Irrigation factors

(1) Total dissolved solids = 509.00
Good quality for irrigation

(2) Conductivity = 1100.000
Good to injurious type of water for irrigation

(3) Concentration of Cl = 1.946
Excellent quality for irrigation

(4) Potential soil salinity = 2.54

(5) Permeability index = 5.72

(6) Percentage of Na = 5.72
unsuitable for irrigation

(7) Sodium absorption ratio = 17.21
Medium-sodium water; suitable only for coarse textured or organic soil with good permeability

(8) Residual sodium carbonate = 50
Safe for normal crop
Valyshko Factor = -1.80

ION exchange calculations (Scholer, 1956)

Positive ion exchange = -1.87
Negative ion exchange = -9.8

Calculation on ionic ratio

The value of H_2SiO_4 is	=	100.00000
Total dry carbonate	=	75.722
Mg-Ca ratio	=	000
Total alkalies	=	5.591
Alkalies-chloride ratio	=	2.873
Na-K ratio	=	35.434
Ca + Mg-alkalies ratio	=	.036
Bicarbonate-chloride ratio	=	1.297
SO_4-Cl ratio	=	.610
Chloride-Bromide ratio	=	1555219.875
Chloride-Carbonate ratio	=	.771

Calculation of Chemical Thermodynamics

Ionic strength of the solution is = .006496
The value of constants A and B are : 5.318999887 e-01 3.321000040e-01

Activity coefficient of the elements

'CA	'	.7116
'MG	'	.7144
'NA	'	.9147
'K	'	.9127
'CL	'	.9127
$'SO_4$	'	.7000

'HCO_3	'	.9174
'SIO_2TOT	'	1.0000
'FE	'	.7116
'FE	'	.4887
'CO_3	'	.7082
'F	'	.9137
'H	'	.9235
'LI	'	.9185
'MN	'	.7116
'NO_3	'	.9127
'OH	'	.9137
'BR	'	.8147

Saturation indices

Ionic activity product for calcite in solution.0000000000
saturation index for calcite = 4.41

Calcite undersaturated in the solution

Ionic activity product for gypsum in solution .000000295
Saturation index for gypsum = -2.89

Gypsum undersaturated in the solution

Data used in this program runs as follows

'CA	'	2,	40.080,6.0
'MG	'	2,	24,3120,,6.5
'NA	'	1,	22.9898,4.0
'K	'	1,	39.1020,3.0
'CL	'	-1,	35.4530,3.0
'SO_4	'	-2,	96.0616,4.0
'HCO_3	'	-1,	61.0173,5.4
'SIO_2TOT	'	0,	60.0848,0.0
'FE	'	2,	55.8470,6.0
'FE	'	3,	55.8470,9.0
'CO_3	'	-2,	60.0094,5.4
'F	'	-1,	18.9984,3.5
'H	'	1,	1.0080,9.0
'LI	'	1,	6.9390,6.0
'MN	'	2,	54.9400,6.0
'NO_3	'	-1,	62.0049,3.0
'OH	'	-1,	17.0074,3.5
'BR	'	-1,	79.9090,4.0

'CA	'	4.000
'MG	'	.001
'NA	'	125.000
'K	'	6.000
'CL	'	69.000
'SO_4	'	57.000
'HCO_3	'	154.000
'SiO_2TOT	'	130.000
'FE	'	.000
'FE	'	.000
'CO_3	'	.000
'F	'	3.000
'H	'	.000
'LI	'	.000
'MN	'	.000
'NO_3	'	.000
'OH	'	.000
'BR	'	.000

1 - Number of datas wanted to be processed. (Optional)
2 - Name of the sample
3 - Unitflag (0-ppm, 1-epm, 2-molality) (Optional)
4 - 5 - Prflat (1 indicates yes for printing and 0 for no to printing) (Optional)
6 - 9 - Cflat (1) to (4) (1 - to call a subroutine and 0 - not to call a subroutine) (Optional)
10 - temperature in degrees
11 - field p^H
12 - lab p^H
13 - Total Dissolved Solids
14 - Conductivity
15 - 32 - Concentration of the species in the order as shown in the Fig. 30.2.

The Effect of Hot Water Effluents from a Thermal Power Plant on the Hydrography and Plankton of River Krishna Near Ibrahimpatnam, Vijaywada (A.P.)–A Case Study

P. AJITHA,*
M.K. DURGA PRASAD*
P.S.R. ANJANEYULU
AND
P. PADMAVATHI

Introduction

There are a few reports on the ecology of the lotic water environments in India. (Chacko and Ganapathi 1949, Ray 1949, Iyengar and Venkatraman 1951, Chakraborthy, Roy and Singh 1959, Philiopose 1959, Roy and Singh and Sehgal 1966, Radha Krishna 1991, Syama Sundar *et al.*, 1991) and a very few reports on the impact of thermal effluents on the hydrology and plankton of rivers in India (Chandra *et al.*, 1985). The present paper attempts to bring out the effect of thermal effluents discharged from Vijaywada thermal power station located at Ibrahimpatnam on the hydrology and plankton of River Krishna.

Physiography of River Krishna and Selection of Sampling Sites

The river Krishna originates near Mahabaleswar at 1,360 m elevation from a water spring. It flows for 1,400 kms and joins Bay of Bengal. Its drainage area is 2,50,948 sq. km of which 26.8% lies in Maharashtra state, 43.8% in Karnataka state and 29.4% in Andhra Pradesh. The river passes through a narrow gorge from Sangameswar to a distance of 130 kms.

The river Krishna at Vijaywada thermal station is having a width of 2.8 kms and a depth of about 12 feet. Fresh water conditions prevail throughout the year there being no traces of tidal influence from Bay of Bengal. Water is highly turbid during flood season. The inflow of water varies from 966 cusecs in June 1990 to 4,22,883 cusecs in August 1990.

The Vijaywada thermal power station is located at Ibrahimpatnam 2 kms away from the river. This thermal power station is having 4 units for generation of electricity. Every day, 190 cusecs of water for unit I, 280 for unit II, and 320 cusecs for unit III and 320 cusecs for unit IV are taken from river Krishna for cooling the condensors of the power plant and this water is again released into the effluent canal. This water picks up high temperature and flows into the effluent canal which traverses for about 1.9 kms and enter river Krishna.

To study the effect of thermal effluents on the hydrology and plankton of river Krishna, three different stations were selected for sampling (Fig. 31.1). This study was taken up from January 1990 to December 1990. Fortnight samples were collected. The first fortnight sampling in every month was collected in the first week at 8 A.M. in the morning. The second sampling was taken in the third week of every month at 3 P.M.

Station I (near the pump house) : This station is located at Vijaywada thermal power station campus near pump house at about a distance of 0.1 kms. This point is located within the effluent canal.

Station II (Effluent canal mouth) : This station is located 0.2 kms away from the shore of the river and located opposite to the entry point of canal where the hot water mixes with the riverine waters. This station is located at a distance of about 1.9 kms from the pump house.

Station III (Ibrahimpatnam ferry station) : This station is located nearer to Ibrahimpatnam shore 0.4 kms up stream from the effluent canal. There is no impact of thermal effluents at this station and pure riverine conditions are maintained all through the year.

Materials and Methods

Data on the physico-chemical parameters like air and surface water temperature, transparency, pH, dissolved oxygen, total alkalinity, total hardness, chlorides, nitrites, iron and phosphates were estimated. (Table 30.1).

Temperature was measured by using an ordinary thermometer graduated from —10°C to 110°C. Hardson's temperature recorder was used for recording the temperature continuously for a week to note the diel variation of temperature. pH was measured by using an Elico pH meter, oxygen dissolved oxygen was estimated by using modified Winklers method suggested by Golterman and Glymo (1969) and total alkalinity, chlorides, nitrites, nitrates, iron and phosphates were estimated by following the methods given in Golterman and Glymo (1969).

Regular quantitative samples of plankton were taken by filtering 100 litres of surface water through a plankton net made of silk bottling cloth no. 25. The samples were identified and counted by the methods recommended by Welch (1948), Edmondson (1971), Pennak (1978), Pontin

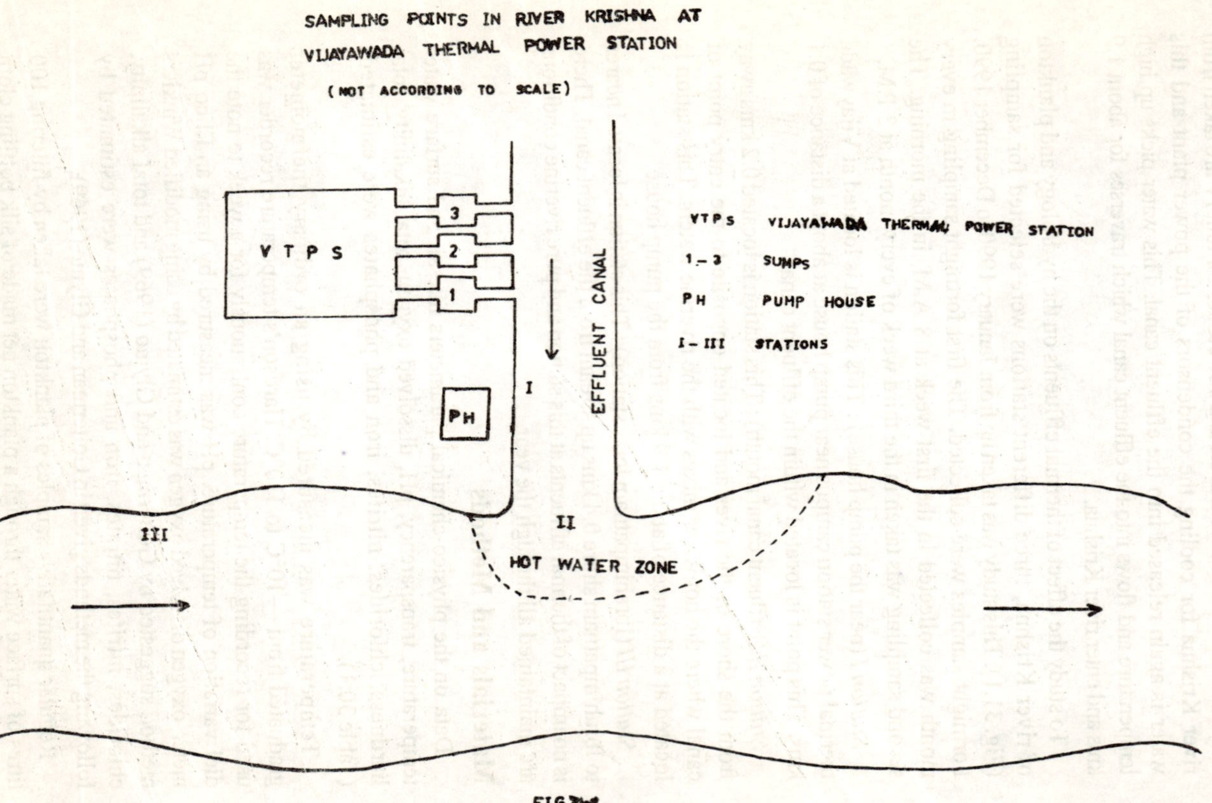
SAMPLING POINTS IN RIVER KRISHNA AT
VIJAYAWADA THERMAL POWER STATION
(NOT ACCORDING TO SCALE)
VTPS VIJAYAWADA THERMAL POWER STATION
1–3 SUMPS
PH PUMP HOUSE
I–III STATIONS
VTPS
3
2
1
PH
I
II
III
EFFLUENT CANAL
HOT WATER ZONE
FIG 31.1

(1978), Fristch (1979), Ranga Reddy (1977), Victor (1985) and Durga Prasad (1991).

Results : (Table 31.1)

Water temperature : The minimum and maximum water temperatures recorded were 34.5°C in July and 42°C in April in Station I, 32°C in October and 38.5°C in April in Station II, 25.9°C in January and 39°C in April in Station III.

Water transparency : The minimum and maximum water transparency recorded were 35 cms in March and October and 60 cms in November and December in station I, 17.3 cms in September and 100 cms in December in Station II, 23 cms in September and 130 cms in April in Station III.

Turbidity : The minimum and maximum values of turbidity were 6 NTU in December and 38 NTU in June in Station I, 7.5 NTU in December and 135 NTU in September in station II, 5 NTU in December and 170 NTU in September in Station III.

Conductivity : The minimum and maximum values ranged between 335 in February and 540 in October in Station I, 340 in February and 540 in October in Station II, 337 in February and 580 in October in Station III.

Dissolved oxygen : The minimum and maximum dissolved oxygen values ranged between 4.8 mg/l in December and a maximum of 9.4 mg/l in November in Station I, 3.4 mg/l in January and 8.4 mg/l in October in Station II and 5 mg/l in January and 9.4 mg/l in February in Station III.

pH : The minimum pH value was 8.3 in January and 9.6 in October in Station I, 8.2 in February and November and 9.6 in October in Station II, 8.3 in January and February and 9.5 in October in Station III.

Total alkalinity : The minimum and maximum values ranged from 103 mg/l in July to 236 mg/l in April in Station I, 105 mg/l in July and 182 mg/l in April in Station II, 63 mg/l in July and 244 mg/l in April in Station III.

Total hardness : The minimum and maximum values ranged between 95 mg/l in December and 165 mg/l in November in Station I, 115 mg/l in August and 197 mg/l in January in Station II, 117 mg/l in July and 187 mg/l in January in Station III.

Chlorides : The minimum and maximum values varied between 15.48 mg/l in July and 52.98 mg/l in May in Station I, 39.8 mg/l in August and November and 56.98 mg/l in June in Station II, 39.8 mg/l in November and 69.97 mg/l in January in Station III.

Nitrates : The minimum and maximum values ranged from 0.0030 mg/l in January to 0.1174 mg/l in March in Station I, 0.0066 mg/l in December to

Table 30.1 : The physico-chemical parameters of river Krishna

MONTH & Station		Temp. A	W	SD	Turb	Cond vity	Do mg/l	pH	TA	CO_3	HCO_3	Ca	Mg	T. Hard	Cholo rides	NO_3	NO_2	PO_4	Iron	T.S.
January	I	28.5	35.0	38	15	397	5.8	8.30	136	36	100	94.9	57.64	153	40.98	0.0030	0.135	0.051	2.050	420
	II	26.5	33.5	35	16	407	3.4	8.30	133	28	105	87.42	110.20	197	44.98	0.0070	0.088	0.072	0.062	470
	III	25.5	25.9	80	8.9	415	5.0	8.30	138	16	122	89.92	97.77	187	69.97	0.0086	0.00695	0.024	0.054	326
February	I	29.5	37.0	46	13	335	7.7	8.35	128	32	106	84.93	65.15	150	43.98	0.0174	0.00825	0.032	0.298	541
	II	30.5	34.5	40	21	340	7.8	8.20	115	20	95	72.43	67.66	140	43.98	0.1584	0.0505	0.0435	0.418	336
	III	27.2	26.5	105	9	337	9.4	8.30	144	22	132	84.92	55.13	140	42.98	0.1678	0.0062	0.0180	0.178	352
March	I	32.3	40.2	35	15	372	7.0	8.35	136	18	118	62.46	92.75	150	46.98	0.1174	0.0342	0.0700	0.556	346
	II	31.0	34.5	38	33	422	7.0	8.30	142	18	124	64.95	102.70	167	49.98	0.1865	0.3698	0.1346	0.648	138
	III	30.0	28.0	68	24	390	7.7	8.50	126	26	102	64.94	90.24	155	47.98	0.1131	0.1044	0.0518	2.330	343
April	I	34.0	42.0	42.5	16	442	7.4	8.50	236	48	188	79.00	48.05	128	47.98	0.0122	0.0096	0.0427	0.356	163
	II	32.5	38.5	45	15	450	7.5	7.8	7.70	8.40	182	40	132	74.90	45.09	120	47.98	0.0068	0.058	0.0548
	III	32.0	39.0	130	14.5	437	7.6	8.50	244	42	202	89.45	54.04	143	51.98	0.0010	0.0043	0.0576	0.216	55
May	I	32.0	41.0	40	18	420	7.4	8.50	129	20	109	77.15	63.34	143	52.98	0.0160	0.0069	0.0430	0.236	371
	II	31.5	36.0	45	15	422	7.5	7.70	130	20	110	119.60	17.87	138	47.98	0.0160	0.0149	0.0465	0.380	275
	III	30.5	28.5	70	14	420	7.9	8.70	138	16	122	87.10	42.89	130	50.98	0.0091	0.0108	0.0225	0.0200	284
June	I	32.7	36.0	40	38	490	6.1	8.65	116	68	96	66.91	74.34	132	50.98	0.0072	0.0135	0.1995	0.200	304
	II	35.2	36.0	33	35	495	6.4	8.80	123	69	55	56.17	83.82	140	56.98	–	0.0440	0.0380	0.272	420
	III	33.5	31.0	40	36	505	7.4	8.80	129	60	69	64.67	73.13	135	53.98	0.0020	0.0104	0.0035	0.140	502
July	I	31.5	34.5	40	11	420	7.0	8.95	103	28	75	62.17	57.33	120	15.48	0.0159	0.0134	0.0416	0.269	252
	II	39.0	33.5	60	12	435	7.6	8.90	105	32	73	67.16	67.85	135	50.96	0.0836	0.2066	0.0677	0.332	315
	III	30.5	27.5	75	10	405	6.4	8.90	63	26	85	64.72	65.33	117	46.98	0.0081	0.0612	0.3840	0.266	314

August	I	31.0	36.0	50	9	460	7.0	9.20	106	8	98	64.42	60.58	125	49.98	0.0180	0.0140	0.0450	0.244	216
	II	29.0	36.5	80	14	450	6.4	9.40	108	12	86	59.42	55.88	115	49.98	0.0289	0.0186	0.4830	0.488	1820
	III	28.0	29.0	90	8	440	8.2	8.9	112	8	104	69.98	57.83	1	49.98	0.0140	0.0110	0.0370	0.136	458
September	I	34.0	40.0	37.5	21	375	7.6	8.4	127	26	102	81.40	76.10	158	36.98	0.0127	0.0083	0.1365	0.184	510
	II	30.5	35.0	17.3	135	460	7.2	8.6	139	18	121	80.61	71.85	152	49.98	0.0289	0.0186	0.4830	0.488	1820
	III	30.0	30.0	23	170	430	8.6	151	32	119	71.91	65.59	170	46.98	0.0649	0.0156	0.1980	0.162	2174	
October	I	34.0	40.0	35.0	19	540	6.4	9.6	168	40	128	79.9	55.10	135	45.98	0.0168	0	0.1110	0.240	450
	II	30.0	32.0	75	18	570	8.4	9.6	178	36	142	54.93	100.00	155	47.98	0.1280	0	0.1140	0.212	210
	III	30.0	28.0	65	16	580	8.6	9.5	196	36	160	69.91	98.05	155	45.98	0.0015	0.0580	0.044	260	
November	I	32.0	38.5	60	8	450	9.4	8.7	132	64	68	89.98	75.10	165	41.98	0.0181	0.0093	0.0440	0.018	246
	II	28.0	36.0	60	18	430	7.8	8.2	124	44	80	84.89·	75.51	160	39.98	0.0253	0.0140	0.362	242	
	III	30.0	28.0	70	10	460	6.4	8.5	138	48	90	85.89	74.51	160	39.98	0.0014	0.4220	0.176	344	
December	I	30.0	39.0	60	6	355	4.8	8.6	160	60	144	44.94	50.06	95	39.98	–	0.0075	0.0205	0.056	174
	II	29.0	35.7	100	7.5	402	7.3	8.3	160	24	137	47.44	72.53	120	40.98	0.0066	0.0054	0.2150	0.149	170
	III	28.0	27.5	88	5	427	7.0	8.6	174	24	150	52.43	72.53	125	43.9	0.0186	0.0392	0.562	102	

0.1865 mg/l in March in Station II, 0.0010 mg/l in April to 0.1678 mg/l in February in Station III.

Nitrites : The nitrate values ranged between zero in October to a maximum of 0.03415 mg/l in March in Station I, zero in October to 0.3698 mg/l in March in Station II, zero in October to 0.0014 mg/l in Station III.

Phosphates : The phosphate values ranged between 0.032 mg/l in February and 0.1995 mg/l in June in Station I, a minimum of 0.027 mg/l in June and 0.1346 mg/l in March in Station II and 0.018 mg/l in February and 0.1297 mg/l in November in Station III.

Iron : The minimum and maximum values of Iron ranged between 0.018 mg/l in November and 2.05 mg/l in January in Station I, 0.062 mg/l in January and 0.608 mg/l in March in Station II, 0.044 mg/l in October and 2.33 mg/l in March in Station III.

Total Solids : The minimum and maximum values of total solids ranged between 163 mg/l in April and 541 mg/l in February in Station I, 128 mg/l in April and 1820 mg/l in September in Station II, 55 mg/l in April and 2174 mg/l in September in Station III.

Biological Parameters

The total plankton (Fig. 31.2) is subjected to significant fluctuations during the period of study at Station I. It is varying from a minimum of 444 individuals per litre in December to a maximum of 1016 individuals per litre in May. At Station II the total number of plankters fluctuated from a minimum of 287 individuals per litre in January to a maximum of 1508 individuals per litre in November whereas at Station III, it ranged from a minimum of 338 individuals per litre in December to a maximum of 749 individuals per litre in March.

The volume of plankton (Fig. 31.3) varied from a minimum of 0.1 ml in October and December to a maximum of 1.3 ml in January, August and September at Station I, 0.1 ml in October to a maximum of 4.2 ml in November at Station II and 0.01 ml in October to a maximum of 2 ml at Station III.

The phytoplankton component (Fig. 31.4) varied from a minimum of 364 numbers per litre in December to a maximum of 961 numbers per litre in May at Station I, 247 numbers per litre in January to 1472 numbers pre litre in November at Station II and 266 numbers per litre in October to a maximum of 684 numbers in April at Station III. Among the phytoplankters, at Station I and II, blue green algae outnumbered the other plankters and massively represented by *Oscillatoria* species, whereas at Station III, *Spirogyra, Fragilaria, Synedra* were the dominant genera in phytoplankton.

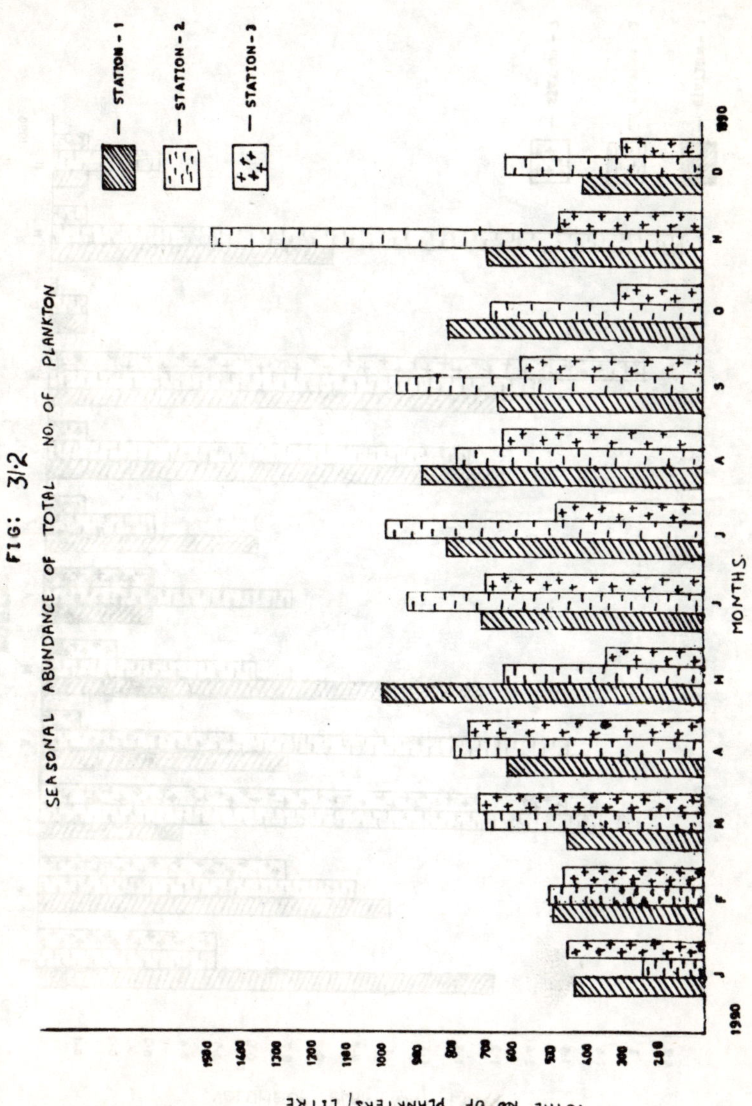
FIG: 3.2
SEASONAL ABUNDANCE OF TOTAL NO. OF PLANKTON
STATION - 1
STATION - 2
STATION - 3
MONTHS
TOTAL NO OF PLANKTERS/ LITRE

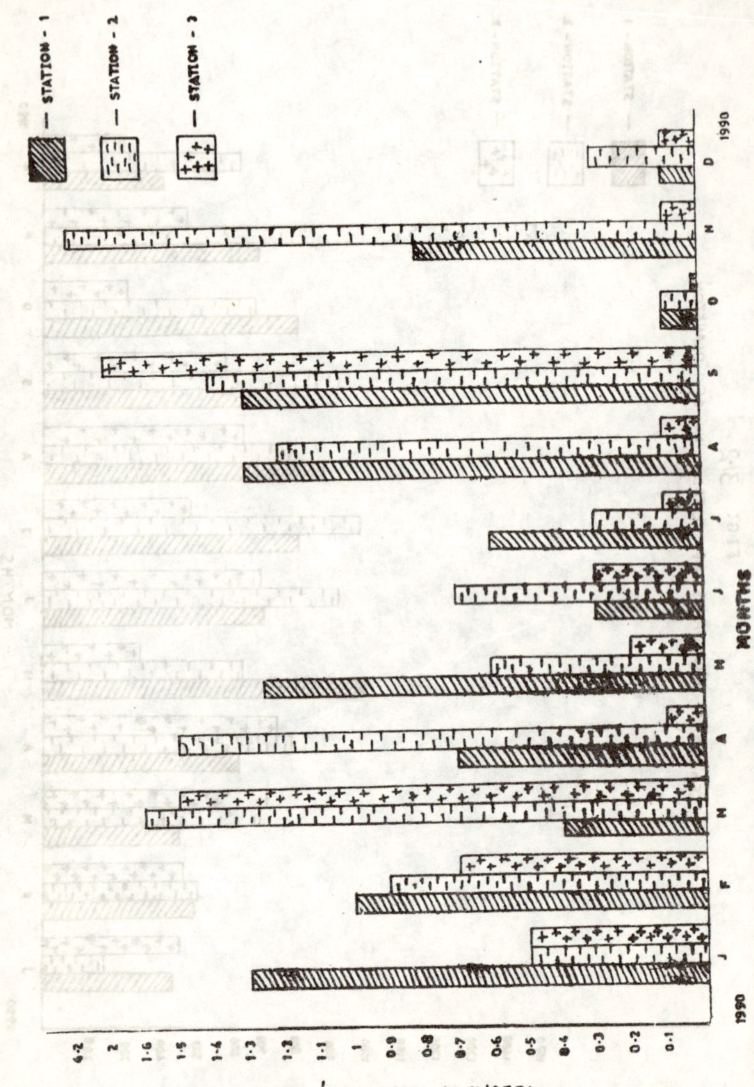
FIG; 31.3
SEASONAL ABUNDANCE OF PLANKTON (BY VOLUME)
STATION - 1
STATION - 2
STATION - 3
VOLUME OF PLANKTON/ LITRE
MONTHS
1990
J
F
M
A
M
J
J
A
S
O
N
D
1990

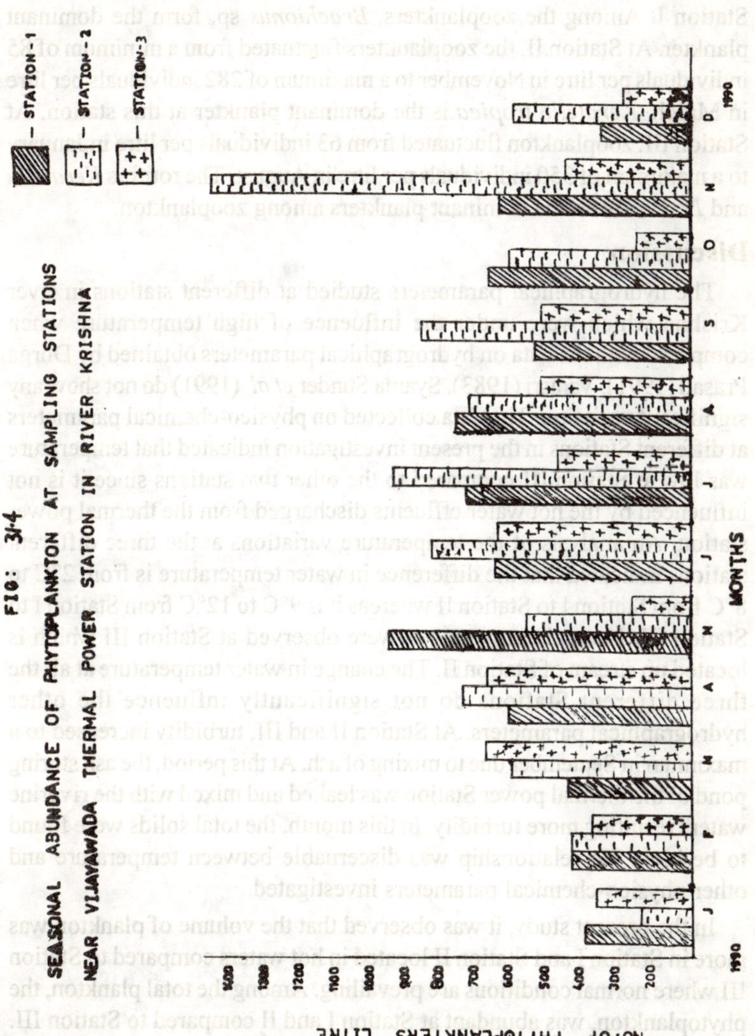
FIG: 31.4
SEASONAL ABUNDANCE OF PHYTOPLANKTON AT SAMPLING STATIONS
NEAR VIJAYAWADA THERMAL POWER STATION IN RIVER KRISHNA
— STATION-1
— STATION-2
— STATION-3
NO. OF PHYTOPLANKTERS LITRE
MONTHS
J F M A M J J A S O N D
1990

The zooplankton (Fig. 31.5) fluctuated from a minimum of 95 individuals per litre in April to a maximum of 352 individuals per litre in October at Station I. Among the zooplankters, *Brachionus* sp. form the dominant plankter. At Station II, the zooplankters fluctuated from a minimum of 85 individuals per litre in November to a maximum of 282 individuals per litre in March. *Keratella tropica* is the dominant plankter at this station. At Station III, zooplankton fluctuated from 63 individuals per litre in January to a maximum of 350 individuals per litre in August. The rotifers *Keratella* and *Lecane* were the dominant plankters among zooplankton.

Discussion

The hydrographical parameters studied at different stations in river Krishna which were under the influence of high temperature when compared with the data on hydrographical parameters obtained by Durga Prasad (1991), Vanisri (1983), Syama Sunder *et al.* (1991) do not show any significant variation. The data collected on physico-chemical parameters at different Stations in the present investigation indicated that temperature was low at Station III compared to the other two stations since it is not influenced by the hot water effluents discharged from the thermal power station. An analysis of the temperature variations at the three different stations indicated that the difference in water temperature is from 2°C to 8°C from Station I to Station II whereas it is 9°C to 12°C from Station I to Station III. Normal temperatures were observed at Station III which is located up stream of Station II. The change in water temperature at all the three different Stations do not significantly influence the other hydrographical parameters. At Station II and III, turbidity increased to a maximum in September due to mixing of ash. At this period, the ash storing pond of the thermal power Station was leaked and mixed with the riverine waters resulting more turbidity. In this month, the total solids were found to be high. No relationship was discernable between temperature and other physico-chemical parameters investigated.

In the present study, it was observed that the volume of plankton was more in Station I and Station II located in hot waters compared to Station III where normal conditions are prevailing. Among the total plankton, the phytoplankton, was abundant at Station I and II compared to Station III. The phytoplankton is primarily constituted by Chlorophyceae, Myxophyceae and Bacillariophyceae. Qualitative analysis of the phytoplankton indicated that there is no substantial variation in its components when compared with the data of Vanisri (*loc. cit*).

The quantitative analysis of the phytoplankton on all the stations in relation to the temperature showed the following results : Blue green

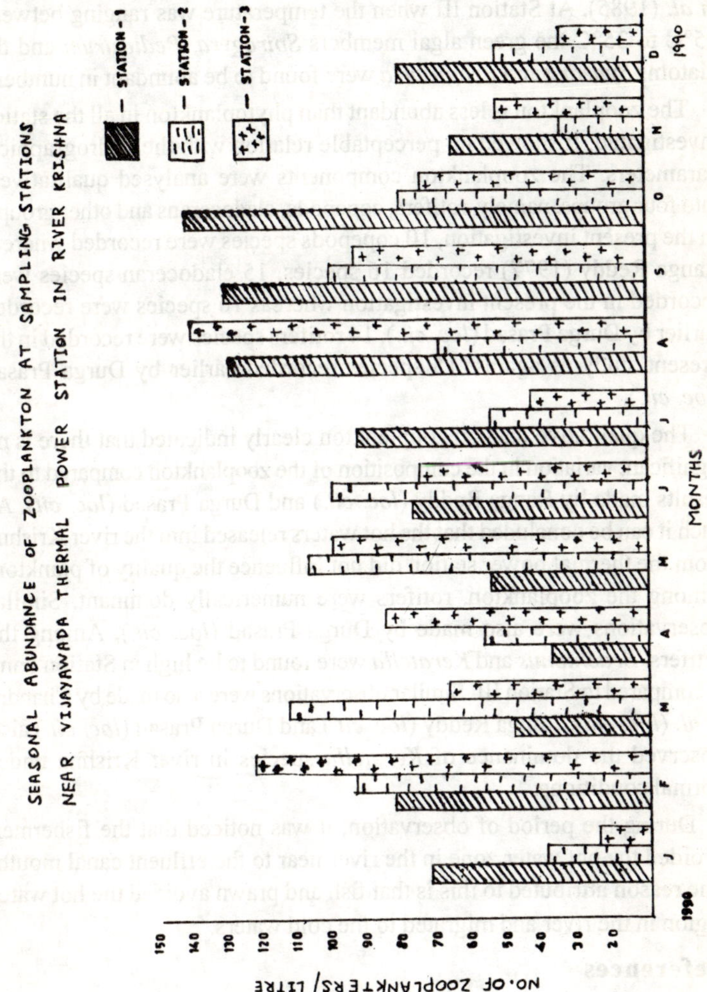
FG: 31.5
SEASONAL ABUNDANCE OF ZOOPLANKTON AT SAMPLING STATIONS
NEAR VIJAYAWADA THERMAL POWER STATION IN RIVER KRISHNA
— STATION-1
— STATION-2
— STATION-3
NO. OF ZOOPLANKTERS/ LITRE
150
140
130
120
110
100
90
80
70
60
50
40
30
20
1990
J
F
M
A
M
J
J
A
S
O
N
D
1990
MONTHS

algal component was dominant at 35°C to 45°C. Green algae and diatoms were dominant at 25°C to 35°C. Similar observations were made by Chandra *et al.* (1985). At Station III when the temperature was ranging between 25°C to 35°C the green algal members *Spirogyra, Pediastrum* and the diatoms *Synedra* and *Fragillaria* were found to be abundant in numbers.

The zooplankton is less abundant than phytoplankton in all the station investigated. It showed no perceptable relation with the hydrographical parameters. The zooplankton components were analysed qualitatively into four groups, namely, rotifers, copepods, cladocerans and other groups. In the present investigation, 10 copepods species were recorded whereas Ranga Reddy (1977) recorded 16 species. 15 cladoceran species were recorded in the present investigation whereas 16 species were recorded earlier by Durga Prasad (*loc. cit.*), 14 rotifers species were recorded in the present study as against 15 species recorded earlier by Durga Prasad (*loc. cit.*).

The qualitative data on zooplankton clearly indicated that there is no significant variation in the composition of the zooplankton compared to the results made by Ranga Reddy (*loc. cit.*) and Durga Prasad (*loc. cit*). As such it can be concluded that the hot waters released into the river Krishna from the thermal power station did not influence the quality of plankton. Among the zooplankton, rotifers were numerically dominant. Similar observations were also made by Durga Prasad (*loc. cit.*). Among the rotifers, *Brachionus* and *Keratella* were found to be high in Station I and II compared to Station III. Similar observations were also made by Chandra *et al.* (*loc. cit.*). Ranga Reddy (*loc. cit.*) and Durga Prasad (*loc. cit.*) also observed the dominance of *Keratella* species in river Krishna under normal conditions.

During the period of observation, it was noticed that the fishermen avoided the hot water zone in the river near to the effluent canal mouth. The reason attributed to this is that fish and prawn avoided the hot water region in the river and migrated to the cold waters.

References

Chacko, P.I and Ganapathi, S.V. (1949) : Some observations on the Adayar river, with special reference to its hydrobiological conditions. Indian Georqr. J. 24.

Chakroborthy, R.D., Roy & Singh (1959) : A quantitative study of the plankton and the physico-chemical conditions of the river Jamna at Allahabad in 1954-55. Indian. J. Fish. 6 : 186-203.

Chandra, K., Singh, D.N. & R.S. Panwar (1985) : Possible pollution problems from wastes of thermal power plants in India a case study of Rihand reservoir, Mirjapur (Uttar Pradesh). Proc. symp. Assess. environ. pollut : 283-94.

Durga Prasad, M.K. & Y. Radha Krishna (1991) : Taxonomy and seasonal fluctuations in Cladocera of river Krishna near Vijaywada, South India. Water ecology pollution and management. Chugh Publications, Allahabad, V. 1 : 172-88.

Edmondson, W.T. (1971) : Manual on methods for the assessment of secondary productivity in freshwaters. I.B.P. Handbook, 17: xxiv + 358.

Edmondson, W.T. (Ed.) (1966) : Freshwater Biology, John Wileysons: 1248. Fritsch F.E. (1931) : Some aspects of the ecology of freshwater algae (with special reference of static waters). J. Ecol., 19 : 233-72.

Golterman, H.L. and Glymo, R.S. (1969) : Methods for chemical analysis of freshwaters, IBP Hand Book No. 8: 172.

Iyengar, M.O.P. and G. Venkatraman (1951) : The ecology and seasonal succession of the river Cooum at Madras with special reference to the Diastomaceae. Jr. of Madras Univ. 21, 140-92.

Pennak, R.W. (1978) : Freshwater invertibrates of the United States. IInd edn. A Wiley Interscience Publication ; 803.

Philipose, M.T. (1959) : Fresh water phytoplankton of Inland Fisheries. Proc. Symp. Algol. ICAR. New Delhi. 272-91.

Pontin, M.R. (1978) : A key to British Freshwater plankton Rotifera, Freshwater Biological Association Scientific Publication No. 38-178.

Radha Krishna, Y. and Y. Ranga Reddy (1976) : Habitat preference in some common fresh water calanoid copepods in the lower deltaic region of the river Krishna, Mem. Soc. Zool. Guntur-1 : 66-69.

Ranga Reddy, Y. (1977) : Studies on systematics and ecology of free-living freshwater copepods of Guntur and its environment (Andhra Pradesh, India), Ph.D. Thesis, Nagarjuna University (Unpubl.).

Ray, P, Singh, S.B. and K.L. Sehgal (1966) : A study of some aspects of ecology of the rivers Ganga and Jamna at Allahabad (U.P.) in 1958-59. Proc. Nat. Acad. Sci. India, 36(3) : 235-72.

Roy, H.K. (1949) : Some potomological aspects of the river Hoogly in relation to Calcutta water supply. Sci. Cult, 14 (13) : 318-26.

Syama Sundar, B., Madhu., G., Srinivas Murthy, K. & V. Mangathayaramma (1991) : Krishna river–its profile and quality. Water ecology pollution and management. Chugh Publications, Allahabad. V. 1 : 189-210.

Vanisri, V. (1983) : Studies on primary productivity in river Krishna and two fresh water bodies in Nagarjuna University Campus, Ph.D. Thesis, Nagarjuna University, India. (Unpubl.)

Victor, N.L. (1985) : Systematics of rotifers from Guntur district (Andhra Pradesh, India), M. Phil Thesis, Nagarjuna University. (Unpubl.)

Vogel, A.E., (1961) : A text book of quantitative inorganic analysis. 3rd edn., Longmans Green and Co., London.

Welch, P.S. (1948) : Limnological methods, Blekiston, Philadelphia: 38.

Concentration of Toxic Trace Elements in Deep Bore Hole Waters at Some Locations in Hyderabad, A.P.

V.K. SAXENA
AND
S. VIJAYA MOHAN

Introduction

The National Geophysical Research Institute, Survey of India, Geological Survey of India, State Bus Complex, Uppal and Nacharam lake are surrounded by several industries, *i.e.*, medicinals, chemicals, dyes, oil and fats, acids, potteries, textiles, metals and confectioneries, etc. Industrial waste contain large amount of inorganic and organic compounds. Some of the toxic inorganic constituents are As, Pb, Cd, Ni, Cr, etc. Water interacts and dissolves the toxic compounds of the waste material and percolates and contaminate the groundwater. In order to properly assess the quality of water, the water quality standards were developed by the United States Environmental Protection Agency (1976), which has established the maximum contaminant level (permissible limit) for toxic trace elements in public water supply. By comparing with these standards the quality and suitability of water can be judged. The present study is an attempt to identify some toxic trace elements and their concentrations in the deep bore hole waters at National Geophysical Research Institute, Survey of India, Geological Survey of India, State Bus Complex, Uppal and Nacharam Lake (Fig. 32.1).

Material and Methods

Water samples were monitored from five different locations (four samples from deep bore holes of 50m and one from a lake) for 12 months from January 1990 to December 1990. For this, 25 water samples were collected from January to December 1990 from all the above locations and analysed for toxic trace elements. Chemical analysis has been done by Atomic Absorption Spectrometer (AAS) (Varian Tectron) and Inductively Coupled Plasma Mass Spectrometer (ICP-MS) (VG Isotopes) at National Geophysical Research Institute. The As, Pb and Cd were analysed by ICP-MS, whereas Ni and Cr by AAS.

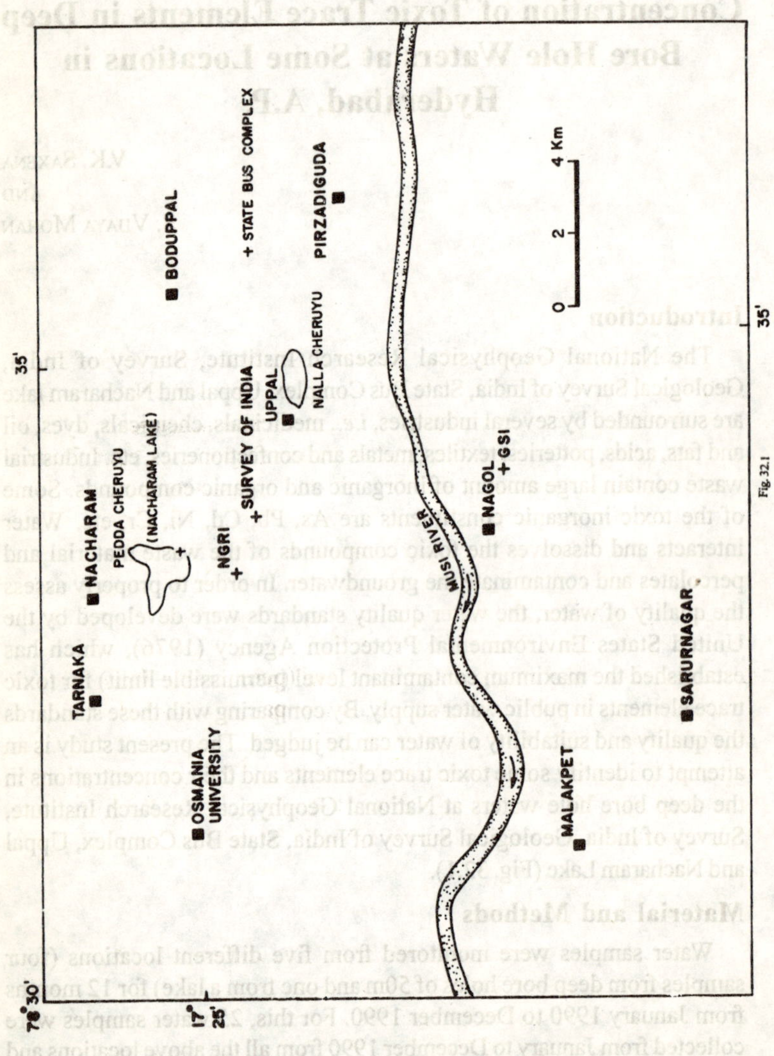
TARNAKA
NACHARAM
PEDDA CHERUYU
(NACHARAM LAKE)
BODUPPAL
OSMANIA UNIVERSITY
NGRI
SURVEY OF INDIA
STATE BUS COMPLEX
UPPAL
NALLA CHERUYU
PIRZADIGUDA
MUSI RIVER
NAGOL
GSI
MALAKPET
SARURNAGAR
0 2 4 Km
78° 30'
17° 25'
35'
35'

Fig. 32.1

Results and Discussion

Arsenic

The weed killers, insecticides and many industrial effluents may contain Arsenic. When these effluents come in contact with fresh water, the result becomes highly problematic. Arsenic is a most powerful chemical pollutant and a very harmful substance for any kind of life system. Arsenic effects the central nervous system and causes polyneuritis, optic neuritis, skin pigmentation, eye congestion, photophobia, abdominal cramps, liver problem and anaemic (Reddy, 1981). The maximum permissible limit of Arsenic in water is 0.1 mg/l (USPHA, 1988). The present water samples have < 0.005 mg/l of Arsenic (Fig. 32.2) and thus all these water samples are within the maximum permissible limit of Arsenic.

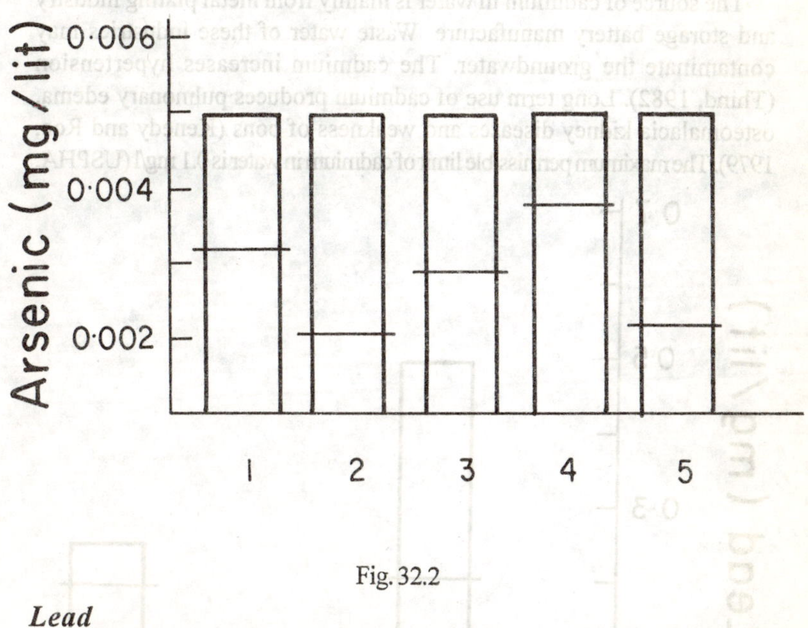

Fig. 32.2

Lead

Lead is considered as one of the most toxic element and very harmful to the human body. Lead accumulates in the bones and soften the tissues and inhibits the formation of haemoglobin leading to anaemia (Sukla and Le Land, 1973). Excess lead in water causes mental retardation, cerebral palsy and damage to optic nerve (Perlstein and Attala, 1976). Presence of lead in water produces anaemia, change of colour of gums, intestinal problem and constipation, muscles weakness, hypertension, nephritis menstrual irregularities, etc. The major source of the lead in groundwater is from the effluents of paints, lead solder, pipe, dyeing and engraving industries (Hammer, 1986). The concentration of lead in present water samples is shown in Fig. 32.3. The maximum permissible limit of Lead in water is 0.1 mg/l (USPHA, 1988). The lead content in National Geophysical Research Institute (NGRI), Survey of India and Geological Survey of India (GSI) is < 0.1 mg/l and is within the permissible limit whereas it varies from 0.1 to 0.5 mg/l in Nacharam lake and 0.1 to 0.25 mg/l in State Bus Complex, Uppal with an average value of 0.2 mg/l at both places. Thus, the lead concentration at Nacharam lake and State Bus Complex, Uppal is more than the permissible limit.

Cadmium

The source of cadmium in water is mainly from metal plating industry and storage battery manufacture. Waste water of these industries may contaminate the groundwater. The cadmium increases hypertension (Thind, 1982). Long term use of cadmium produces pulmonary edema, osteomalacia kidney diseases and weakness of bons (Kenedy and Roe, 1979). The maximum permissible limit of cadmium in water is 0.1 mg/l (USPHA,

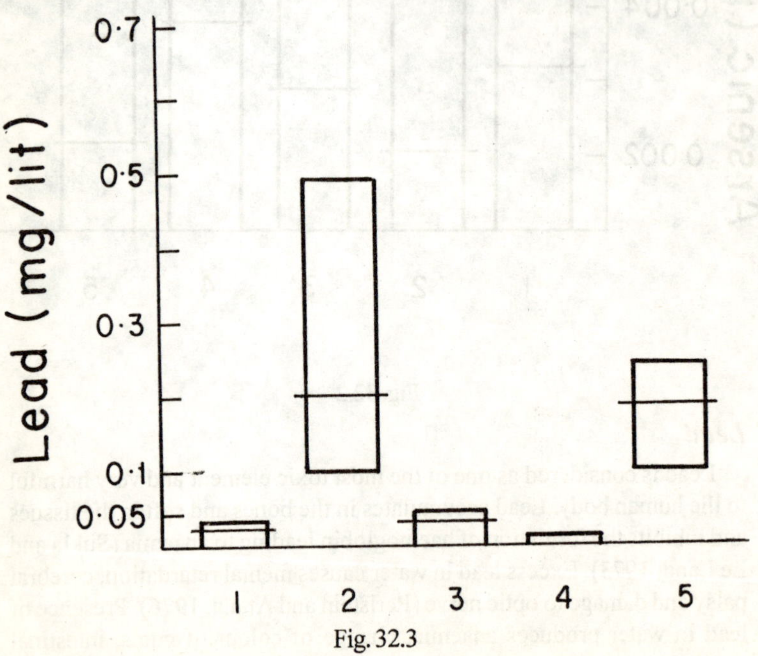

Fig. 32.3

1988). The concentration of Cd varies from 0.1 to 0.32 mg/l with an average value of 0.21 mg/l in Nacharam lake and from 0.1 to 0.24 mg/l with an average value of 0.19 mg/l at State Bus Complex, Uppal. This shows cadmium contamination in the waters of Nacharam lake and the State Bus Complex, Uppal. The cadmium content in the waters of NGRI, Survey of India and GSI is < 0.1 mg/l and it is less than the maximum permissible limit.

Nickel

The main source of nickel in water is from the effluents of cement, cable, metal and chemical industries. Nickel compounds in common form are relatively less soluble and not much toxic but the gaseous nickel carbonyl is highly toxic. Nickel content in water when exceeds 0.1 mg/l causes kidney diseases, digestive disturbances and hypertension (Fitter, 1980). The nickel content in the water samples at NGRI, Survey of India and GSI is < 0.02 mg/l. Thus, these waters are free from nickel pollution. The nickel concentration is much higher (0.1 to 0.2 mg/l) in Nacharam lake and 0.1 to 0.18 mg/l in Uppal Bus Complex (Fig. 32.5) and it indicates higher nickel contamination.

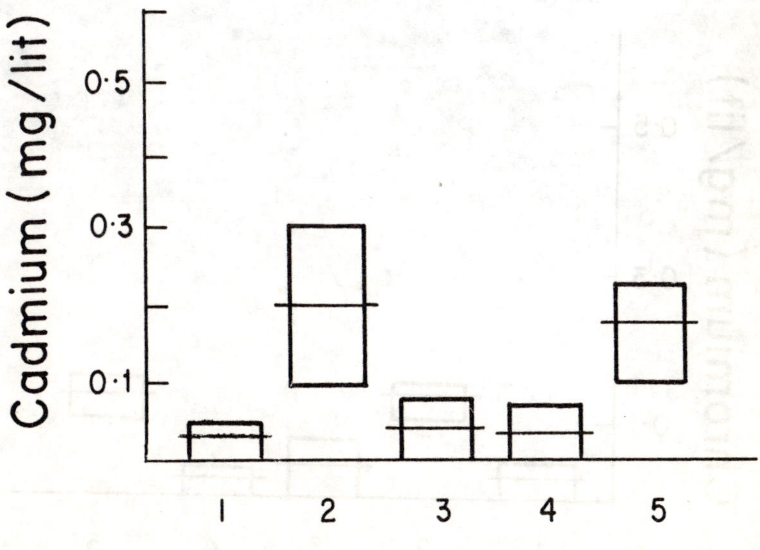

Fig. 32.4

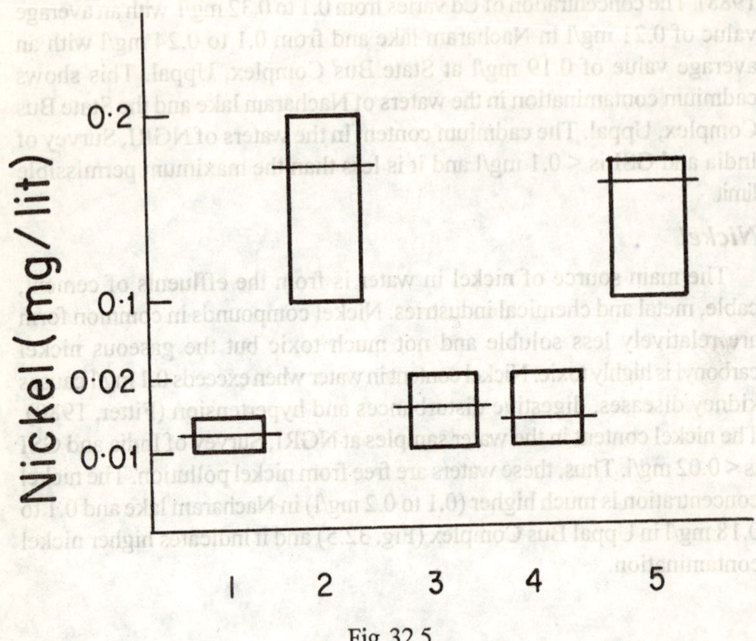

Fig. 32.5

Fig. 32.6

Chromium

The main source of chromium in groundwater is the effluents of metals, chemicals, dyes, potteries and textile industries. Chromium in water is a powerful pollutant. It reduces the resistance of the body. It produces low blood pressure and gastrointestinal disturbances (Reddy, 1981). The maximum permissible limit of chromium in water is 0.1 mg/l (APHA, 1985). The present study indicates that the Cr content at NGRI, Survey of India and GSI is < 0.1mg/l (Fig. 32.6) and it is within the permissible limit. The Cr content at Nacharam lake and Uppal Bus Complex, ranges from 0.1 to 0.15 mg/l with an average value of 0.14 mg/l and it is more than the permissible limit.

Acknowledgement

Thanks are due to officers of Geological Survey of India, Survey of India and State Bus Complex, Uppal for providing the facilities to monitor the water samples. Prof. D. Gupta Sarma, Director, National Geophysical Research Institute, Hyderabad is gratefully acknowledged for his kind permission to publish this paper.

References

American Public Health Association (APHA) Standard methods for the examination of water and waste water, 16th Ed. 1200 p.

Fetter, C.W., (1980) Applied Hydrogeology. Columbus, Ohio, Merrill Publishing Co., 448 p.

Hammer, M.J., (1986) Water and waste water technology, New York, John, Wiley and Sons, 550 p.

Kenedy, G. and Roe, F.J.C., (1979) Cadmium toxicology, Lancet, 1 pp. 1206-07.

Perlstein, N.A. and Attala, R., (1976) Neurologic sequel of plumbism in children, Clinical Pediatics, V. 5, pp. 292-98.

Reddy, K.S.N., (1981) The essentials of forensic medicine and toxicology. 5th Ed. Sugune, Hyderabad.

Sukla, S.S. and Le Land, H.V., (1973) Heavy metals in water, J. Water pollution federation, V. 45, pp. 1319-31.

Thind, G.S., (1982) Role of toxic element in human on experimental conditions, J. Air and Water Pollution Control, Asso., V. 22, pp. 267-70.

United States Environmental Protection Agency (1976), Quality criteria for water, U.S. Govt. Printing Office, Washington, 256 p.

United States Public Health Association, (USPHA) International standards for drinking water. U.S. Govt. Printing Office, Washington, pp. 75-88.

33

Effects of Steel Plant Effluent on Seed Germination and Seedling Characteristics of *Phaseolus Vulgaris* L.

ARCHANA SHARMA
AND
M.L. NAIK

Introduction

Disposal of industrial effluent in to the environment requires before hand knowledge of its quality and its effects on the potentially exposed biota. Based on this idea, studies have been made to evaluate the effects of liquid effluents from industries on seed germination and seedling characteristics (Jerath and Sahai 1982, Sahai *et al.* 1983, Singh *et al.* 1985, Sisodia and Beid 1985, Singh and Mishra 1987, and Neelam and Sahai 1988). However, despite the fact that the effluent from one of the largest integrated steel mills in India, Bhilai Steel Plant (BSP), Bhilai is being used for irrigation since about more than 20 years, very few such studies (Sharma and Naik 1989, 1990a, b, c) have been made on effects of these effluents on plants or soil. The present studies have, therefore, been made to investigate the quality and effects of undiluted effluent from BSP, on seed germination and seedling characteristics of *Phaselous vulgaris* L.

Material and Method

The effluent samples were collected from about 2 km down-stream from the steel plant area. Analyses of effluents were made by following the Standard Methods (APHA-AWWA-WPCF 1975). Other parameters were determined as follows : germination value (Gzabator 1962), speed of germination index (Kamal and Sinha 1974), germination relative index (Zur 1966) and mineral nutrients (Jackson 1973).

Results and Discussions

Physico-chemical characteristics of effluent are given in Table 33.1. Many of these parameters were found to be beyond the limits prescribed by ISI (IS : 3307, 1964). The effects of steel plant effluent on seed germination and seedling characteristics are shown in Table 33.2. In the seeds of *P. vulgaris*, when germinated with effluent, the parameters investigated presently, were affected significantly. Germination percentage, germination value, speed of germination index and germination relative index decreased by about 12%, 27%, 30% and 46% respectively, but the water absorption capacity increased by about 40%. The steel plant effluent grown seedlings also experienced generally significant effect, with about 25% increase in the fresh weight but marginal decrease in the dry weight besides the changes in moisture percentage, magnesium concentration and phosphate. There was a very significant decrease of 49.79% of calcium, under treated condition. Iron concentration increased in treated seedlings by 12.34%. The concentration of total nitrogen and crude protein also exhibited an increase in treated seedlings but very insignificantly.

Table 33.1: Mean value for physico-chemical characteristics of BSP effluent

	Parameters		**Mean ± SD*
1.	pH		7.93 ± 0.36
2.	Total alkalinity	mg/l	93.2 ± 30.1
3.	Chloride	mg/l	57.15 ± 15.75
4.	Phosphate	mg/l	1.045 ± 1.093
5.	Sulphate	mg/l	4407 ± 1993
6.	Nitrite-nitrogen	mg/l	8.41 ± 5.07
7.	Nitrate-nitrogen	mg/l	2804 ± 1400
8.	Ammonia-nitrogen	mg/l	6583 ± 3772
9.	Iron	mg/l	5351 ± 6150
10.	Hardness as Ca	mg/l	173 ± 711
11.	Hardness as $CaCO_3$	mg/l	831 ± 520
12.	Total hardness as $CaCO_3$	mg/l	831 ± 700
13.	Magnesium	mg/l	98 ± 50
14.	Chemical oxygen demand	mg/l	558 ± 454
15.	Phenol	mg/l	46.57 ± 83.96

* 35 replicates.

Present investigation indicates that BSP effluent has significant effects on most of the seed germination and seedling characteristics. These effects are, however, due to undiluted effluent and thus appear to be far less drastic than the magnitude of effects reported for even diluted effluents from some other industries (Jerath and Sahai 1982, Sahai *et al.* 1985, Sisodia and Bedi 1985, Singh and Mishra 1987, Neelam and Sahai 1988). The presently investigated effluent although appears to be less injurious to plants as compared to the effects of effluents from some other industries, the continuation of irrigation with this effluent does not appear to be safe and should be treated more, to make it suitable for irrigation.

Table 33.2 : Effects of BSP effluent on seed germination and seedling characteristics of *Phaseolus vulgaris* L.

	Parameters	*Control (± SD)*	*Treated (± SD)*	*% change in treated over control*	*"t" values*
1.	Germination percentage	85 (0.764)	75 (1.627)	-11.76	10.61*
2.	Germination value	64 (0.764)	47 (1.627)	-26.56	17.75*
3.	Speed of germination index	340 (2.517)	238 (0.577)	-30.00	68.42*
4.	Germination relative index	41 (0.577)	22 (1.527)	-46.34	20.15*
5.	Water absorption capacity of seeds/g/day	1.015 (0.011)	1.422 (0.178)	+40.10	3.96*
6.	Fresh weight of seedling/g seed	2.333 (0.060)	2.914 (0.015)	+24.90	3.92*
7.	Dry weight of seedling/g seed	1.053 (0.0006)	1.052 (0.0006)	-0.09	2.12
8.	Moisture percentage of seedlings	54.88 (0.061)	52.07 (0.059)	-5.12	57.31*
9.	Calcium mg/100g	9.56 (0.062)	4.80 (0.003)	-49.79	132.89*
10.	Magnesium mg/100g	0.717 (0.018)	0.672 (0.003)	-6.28	429*
11.	Phosphate mg/g	3.713 (0.013)	3.581 (0.025)	-3.56	8.09*
12.	Iron mg/g	0.462 (0.021)	0.519 (0.001)	+12.34	4.83*
13.	Total nitrogen %	0.161 (0.026)	0.162 (0.015)	+0.62	0.06
14.	Crude protein %	1.005 (0.163)	1.013 (0.095)	+ 0.80	0.07

* Significant at 5% level.

Acknowledgements

The authors are thankful to the MAPCOST for providing financial assistance through project code env 96/87.

References

APHA-AWWA-WPCF, (1975) *Standard Methods for the examination of Water and Waste water,* 14th edition. Amer. Publ. Health Assoc. Inc. New York.

I.S.I. 3307–(1964). Tolerance limits for industrial effluents discharged on land for irrigation purpose. Indian Standard Institution, New Delhi.

Jackson, M.L., (1973) Soil Chemical Analysis. Prentice-Hall of India Private Limited, New Delhi.

Jerath, N. and Sahai, R., (1982) Effect of Fertilizer factory effluent on seed germination and seedling growth of maize. Indian J. Ecol. 9(2) : 330-34.

Kamal and Sinha, A.B., (1974) Studies on Aspergilli from soils of Gorakhpur IV. Effect of culture filterates of certain Aspergilli on germination of *Abelmoschus esculentus* seeds. Trop. Ecol. 15 (1 & 2) : 11-15.

Neelam and Sahai, R., (1988) Effect of fertilizer factory effluent on seed germination, seedling growth, pigment content and biomass of *Sesamuum indicum* Linn. J. Environ. Biol. 9(1) : 45-50.

Sahai, R., Jasbeen, S.K. and Saxena, P.K., (1983) Effect of distillery waste on seed germination, seedling growth and pigment content of Rice. Indian J. Ecol. 10(1) : 7-10.

Sharma, Archana and Naik, M.L., (1989) Effluent from Bhilai Steel Plant and its effect on black cotton soil and plant species. *Abelmoschus esculentus* (Linn). Moench. Biome. 4(2) : 100-16.

Sharma, Archana and Naik, M.L., (1990a) Characteristics of Bhilai Steel Plant effluent and its effect on soil and plant species (*Crotalaria juncea* L.) *J. NATCON.* 2(1) : 53-161.

Sharma, Archana and Naik, M.L., (1990b) Treated effluent from Bhilai Steel Plant and its effects on plants, J. Env. Sci., 3(1) : 17-24.

Sharma, Archana and Naik, M.L., (1990c) Steel Mill effluents and its effects on soil and plants. Environmental series IV, Environ Concern and Tissue Injury, 15-26.

Singh, D.K., Kumar, Dinesh and Singh , V.P., (1985) Studies on pollutional effects of sugar mill and distillery effluent on seed germination and seedling growth of three varieties of Rice. J. Environ. Biol., 6(1) : 31-35.

Singh, K.K., and Mishra, L.C., (1987) Effects of fertilizer factory effluent on soil and crop productivity. Water, Air and Soil Pollution, 33 : 309-20.

Sisodia, G.S. and Bedi, S.J. (1985) Impact of chemical industry effluent on seed germination and early growth performance of wheat. Indian J. Eco. 12(2) : 189-92.

Zur, B., (1966) Ostomic control of the matrix soil water potential I Soil Plant System. Soil Sci. 102 : 394-98.

34

Groundwater Pollution in Some Parts of Visakhapatnam, India : A Case Study

N. Someswara Rao
T.N.V. Venkateswara Rao,
D.R.R. Sarma, K.G. Naidu
and
M.V. Venkata Rao

Introduction

Visakhapatnam (17°43′N latitude and 83°17′ longitude (Rao, 1971) is located on the east coast of India. The physiography of Visakhapatnam city (Fig. 34.1) which appears like a small basin is surrounded by the Yarada hill popularly known as Dolphin's nose on the southern side and Kailasa hill on the north, with the Bay of Bengal forming the eastern limit. The coastal line runs from north-east of south-west over a distance of six kilometres. On the west, there is an extensive tidal basin called Uppteru now under reclamation.

The residents of Ramakrishnapuram (R.K. Puram) area are mainly depending on groundwater for their drinking and other domestic requirements for the last several years as there was no municipal water supply. The residents of this area observed a sudden change in the quality of water in the summer, 1985.

According to the residents, the water was normal till March, 1985 and latter became hard and saline resembling sea water. So, they were not able to drink the water or use it even for any other domestic purpose. Hence, the aim of the present study is set to discuss the source and the cause for the sudden increase in hardness and salinity in R.K. Puram area.

Material and Methods

Forty-three groundwater samples were collected in and around Ramakrishnapuram area in clean polyethylene bottles. The samples were preserved by adopting standard techniques (APHA, 1985). Various chemical parameters were analysed using standard methods (APHA, 1985, Golterman *et al.*, 1978, Vogel 1978 and Jan and Young, 1978).

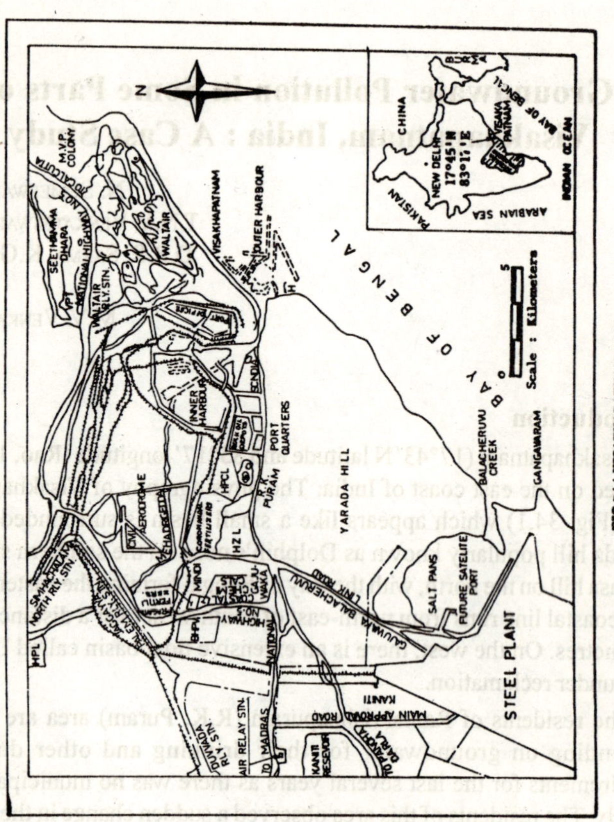

FIG. 34-1 VISAKHAPATNAM MAP

Simulation studies were carried out in the laboratory to confirm the high hardness values. For these studies, soil samples were collected from Jawaharnagar area. The characteristics of the soil in both the areas are similar. A column was prepared with the soil dried at 100°C. Sodium chloride solutions of known concentrations were passed through the column at a slow rate and the leachates were collected dropwise. This was repeated with sodium chloride solution of different concentrations and columns with fresh sample of soil each time, keeping constant the other experimental conditions like quantity of soil, its length in the column and the rate of the collection of the leachates, etc.

Results and Discussion

The results of periodical monitoring of groundwater in R.K. Puram and surrounding areas of Visakhapatnam city during June, 1985 and March, 1988 are shown in Tables 34.1 and 34.2. The changes in hardness and salinity of groundwater with time in R.K. Puram and other surrounding areas are presented in Tables 34.3 and 34.4 respectively.

As can be seen from Table 34.1, the hardness and salinity in the groundwaters of R.K. Puram area are very high whereas in the surrounding areas they are relatively normal. After examining the area thoroughly the following two possibilities have been considered to explain the sudden increase in hardness and salinity of groundwater.

(*i*) Presence of faults in the sub-surface of R.K. Puram area following the sea water to enter the aquifers.

(*ii*) Sea water intrusion into the surrounding aquifers.

The first possibility is ruled out for the following reasons.

(*a*) The suitability of the area for the construction of huge naphtha storage tanks was tested during 1984 by the refinery authorities. If any faults are present they would not have constructed the tanks in that place.

(*b*) Periodical monitoring of groundwater in this area proved that the hardness and salinity have been decreasing (Tables 34.1 & 34.2). If it were due to some faults in the sub-surface the salinity and hardness of groundwater would not have decreased with time, because of the continuous inflow of sea water.

Table 34.1 : Analysis of Water Samples Collected from Ramakrishnapuram Area (20.6.85)

S. No.	*pH*	*E.C.*	*T.D.S.*	*Hardness*	*Calcium*	*Magnesium*	*Chloride*
1.	7.16	1.70	1,906	1,101	892	209	235
2.	7.22	14.27	29,860	21,548	18,450	3,098	5,554
3.	7.17	18.42	36,550	27,534	21,918	5,616	7,290
4.	7.41	28.50	56,150	39,505	32,142	7,363	11,638
5.	6.98	16.00	19,037	13,168	9,698	3,470	3,471
6.	7.08	20.45	44,728	32,438	23,963	8,475	9,524
7.	7.02	2.42	1,829	825	607	218	362
8.	6.99	3.82	4,658	2,442	1,806	636	1,375
9.	7.02	5.06	7,500	4,788	3,538	1,250	1,469
10.	6.88	6.42	10,616	6,816	5,029	1,787	2,614
11.	7.12	6.63	15,289	6,227	5,121	1,806	7,182
12.	6.98	5.84	3,684	2,048	1,512	536	812
13.	7.02	5.99	3,594	2,124	1,566	558	665
14.	7.05	4.82	3,142	1,616	1,192	424	802
15.	7.10	2.48	2,197	982	724	258	484
16.	7.08	1.32	1,436	498	362	136	228
17.	7.07	1.72	1,174	464	342	122	190
18.	7.05	1.07	785	253	186	67	93
19.	7.12	1.33	980	350	254	96	138
20.	7.10	5.45	9,683	7,183	5,318	1,865	1,475
21.	7.01	8.70	14,408	10,883	8,046	2,837	2,423
22.	7.01	2.92	3,682	1,994	1,449	545	1,027
23.	7.17	1.53	1,220	497	362	135	174
24.	7.07	2.73	1,879	788	581	207	408
25.	7.12	2.19	1,638	632	466	166	392
26.	7.01	2.32	1,826	754	552	202	491
27.	7.02	2.01	1,633	658	488	170	318
28.	7.28	2.68	1,349	482	352	130	337
29.	7.36	2.92	1,568	589	434	155	329
30.	7.24	1.24	1,054	362	268	94	182

Table 34.1 (Contd.)

S. No.	pH	E.C	T.D.S.	Hardness	Calcium	Magnesium	Chloride
31.	7.12	1.26	1,047	368	271	97	193
32.	7.02	0.98	918	312	231	81	161
33.	7.18	0.82	623	156	112	44	83
34.	7.15	2.68	1,930	778	573	205	479
35.	6.98	0.97	988	342	252	90	157
36.	7.02	1.12	1,388	384	286	98	303
37.	7.08	1.06	1,282	416	306	110	254
38.	6.87	1.39	1,642	575	422	153	398
39.	7.08	1.16	1,184	388	285	103	150
40.	7.02	0.54	648	146	106	40	81
41.	6.99	0.82	720	218	159	59	112
42.	6.94	0.80	718	217	156	61	109
43.	7.04	0.71	704	208	151	57	92

S. No.	Sulphate	Phosphate	Nitrite	Nitrate	Alkalinity	Fluoride
1.	184	20	0.042	14	297	0.69
2.	540	20	0.131	22	305	0.82
3.	870	23	0.129	15	294	0.74
4.	2,028	20	0.082	16	259	0.68
5.	660	16	0.048	17	248	0.52
6.	1,648	20	0.041	18	282	0.64
7.	260	18	0.043	15	272	0.48
8.	473	17	0.057	16	256	0.51
9.	528	21	0.022	15	238	0.48
10.	614	22	0.051	18	225	0.52
11.	625	21	0.014	27	262	0.46
12.	413	20	0.042	24	216	0.42
13.	424	17	0.041	25	225	0.42
14.	351	16	0.051	18	225	0.52
15.	248	10	0.002	12	284	0.67

Table 34.1 (Contd.)

S. No.	Sulphate	Phosphate	Nitrite	Nitrate	Alkalinity	Fluoride
16.	189	10	0.003	14	268	0.67
17.	164	11	0.079	19	161	0.69
18.	87	10	0.007	12	252	0.49
19.	122	9	0.041	10	275	0.66
20.	674	11	0.053	15	262	0.52
21.	829	10	0.058	14	233	0.58
22.	361	12	0.032	13	224	0.59
23.	132	24	0.038	15	285	0.61
24.	302	22	0.044	18	271	0.41
25.	239	19	0.041	17	289	0.59
26.	308	17	0.035	19	248	0.54
27.	241	18	0.009	18	255	0.51
28.	148	20	0.095	21	278	0.54
29.	159	21	0.092	24	294	0.56
30.	128	19	0.098	21	265	0.65
31.	129	7	0.004	10	278	1.02
32.	112	8	0.006	9	255	0.98
33.	76	7	0.001	8	284	0.52
34.	328	12	0.002	17	271	0.51
35.	216	9	0.001	12	228	0.61
36.	262	11	0.001	14	239	0.61
37.	289	10	0.002	15	248	0.36
38.	312	14	0.002	16	221	0.32
39.	219	12	0.002	18	253	0.61
40.	54	10	0.003	15	237	0.64
41.	125	10	0.002	13	224	0.63
42.	121	9	0.001	14	188	0.39
43.	104	10	0.001	13	239	0.43

Note : E.C. is expressed in m. mhos, remaining values are expressed in mg/l.

Table 34.2 : Analysis of Water Samples Collected from Ramakrishnapuram Area (30.12.87)

S. No.	pH	E.C.	T.D.S.	Hardness	Calcium	Magnesium	Chloride
1.	7.26	0.79	925	310	243	67	128
2.	7.18	1.47	1,252	433	318	115	206
3.	6.99	3.47	3,682	1,571	1,064	507	731
4.	7.18	3.71	4,396	2,193	1,556	637	937
5.	7.15	2.40	2,348	1,015	765	250	408
6.	7.24	2.81	2,817	1,084	786	298	742
7.	6.98	1.33	1,528	316	245	71	168
8.	7.09	1.21	924	382	289	93	302
9.	7.18	2.10	1,437	684	521	163	317
10.	7.28	2.45	2,098	883	668	215	350
11.	7.08	2.01	2,416	792	602	190	693
12.	6.85	1.51	1,324	556	428	128	133
13.	7.18	1.24	1,882	325	308	117	416
14.	7.14	1.08	1,352	498	287	111	262
15.	7.08	0.92	1,094	304	228	76	178
16.	7.07	0.72	947	260	198	62	102
17.	7.08	7.53	7,972	3,925	2,993	932	2,268
18.	7.07	1.02	1,416	458	342	116	102
19.	7.42	0.89	976	296	223	73	98
20.	7.22	2.65	2,184	892	678	214	329
21.	7.15	2.86	2,428	993	752	241	516
22.	7.12	1.53	1,509	562	423	139	238
23.	7.21	1.18	1,247	364	272	92	122
24.	7.19	1.32	1,416	416	309	107	214
25.	7.18	1.18	1,312	392	282	110	219
26.	6.99	1.08	1,208	397	293	104	216
27.	7.18	1.24	1,354	406	304	102	208
28.	7.32	9.42	6,889	3,403	2,434	969	1,855
29.	7.01	7.84	5,778	2,996	2,248	748	1,080

Table 34.2 (Contd.)

S. No.	*pH*	*E.C.*	*T.D.S.*	*Hardness*	*Calcium*	*Magnesium*	*Chloride*
30.	7.28	4.18	3,406	1,298	967	331	867
31.	7.25	0.89	982	302	224	78	168
32.	7.14	0.83	968	289	214	75	125
33.	6.98	0.53	724	198	142	56	97
34.	7.05	2.62	201	4762	564	198	501
35.	6.94	0.89	978	260	193	67	109
36.	7.14	1.02	1,288	361	269	92	135
37.	7.11	0.78	982	218	162	56	118
38.	7.04	0.84	1,042	319	238	81	216
39.	7.14	0.98	1,184	393	285	108	143
40.	6.99	0.95	1,138	391	279	112	159
41.	6.98	1.28	1,456	491	264	227	287
42.	7.04	0.85	1,018	282	209	73	153
43.	7.08	0.94	1,104	298	223	75	165

S. No.	*Sulphate*	*Phosphate*	*Nitrite*	*Nitrate*	*Alkalinity*	*Fluoride*
1.	65	25	0.066	17	312	0.78
2.	175	23	0.028	25	292	0.68
3.	516	27	0.249	24	244	0.56
4.	628	24	0.104	22	324	0.62
5.	217	20	0.062	27	313	0.42
6.	226	24	0.068	22	308	0.68
7.	68	23	0.075	16	249	0.52
8.	79	21	0.108	18	271	0.55
9.	104	28	0.052	19	216	0.56
10.	208	27	0.094	22	324	0.54
11.	219	28	0.062	31	336	0.49
12.	172	26	0.077	25	235	0.48
13.	184	21	0.090	29	340	0.44

Table 34.2 (Contd.)

S. No.	*Sulphate*	*Phosphate*	*Nitrite*	*Nitrate*	*Alkalinity*	*Fluoride*
14.	152	20	0.082	31	340	0.44
15.	127	12	0.005	16	262	0.78
16.	98	14	0.004	17	248	0.60
17.	497	16	0.115	24	254	0.78
18.	156	16	0.012	18	251	0.52
19.	98	15	0.098	15	309	0.74
20.	228	17	0.099	18	291	0.62
21.	256	18	0.102	16	289	0.67
22.	183	17	0.092	17	278	0.66
23.	115	27	0.091	16	289	0.67
24.	213	26	0.094	24	271	0.52
25.	139	25	0.095	21	268	0.68
26.	134	22	0.087	24	229	0.61
27.	136	23	0.038	26	274	0.63
28.	816	28	0.210	29	324	0.63
29.	799	27	0.202	32	292	0.69
30.	361	25	0.208	28	288	0.72
31.	94	14	0.034	12	281	1.24
32.	91	14	0.036	13	294	1.25
33.	86	9	0.003	12	238	0.062
34.	316	16	0.006	24	260	0.59
35.	177	14	0.004	21	280	0.48
36.	224	16	0.003	17	277	0.74
37.	124	15	0.004	21	280	0.48
38.	153	18	0.003	22	265	0.42
39.	224	17	0.005	24	285	0.69
40.	219	15	0.008	16	258	0.78
41.	288	16	0.007	16	242	0.71
42.	148	14	0.003	17	258	0.52
43.	157	14	0.004	18	264	0.55

E.C. is expressed in m. mhos, remaining values are expressed in mg/l.

Particulars of Water Samples Collected from Ramakrishnapuram and Surrounding Areas

Sample No.	*Source*	*Particulars*
1.	Bore-well	Near St. Joseph's School, R.K. Puram
2.	Bore-well	Near H/o P. Appanna; H.No. 61-8-1/1, R.K. Puram
3.	Bore-well	Inside the H. No. 61-1-53; R.K. Puram
4.	Bore-well	Near H. No. 61-1-53; R.K. Puram
5.	Bore-well	Near H/o A. Rajarao; H. No. 61-2-8; R.K. Puram
6.	Bore-well	Opp. to H. No. 61-8-8; R.K. Puram
7.	Bore-well	Adjacent to HPCL Compound wall; R.K. Puram
8.	Bore-well	Near H. No. 60-1-82; R.K. Puram
9.	Bore-well	Near H. No. 59-12-59; Malkapuram
10.	Bore-well	H/o K. Apparao; H. No. 59-1-35; Malkapuram
11.	Well	H.p B. Konda, H. No. 59-1-35; Malkapuram
12.	Bore-well	Near No. 59-11-49; Malkapuram
13.	Bore-well	Near H/o D. Narayana; H. No. 59-14-8; Malkapuram
14.	Bore-well	Opp. to H/o P.S. Raju; H. No. 59-14-8; Malkapuram
15.	Well	Which is supplied to Port quarters; Malkapuram
16.	Well	Old pump house of Port; Malkapuram
17.	Bore-well	Opp. to Shipyard New Colony; Malkapuram
18.	Public well	Near Mahalakshmi Theatre, Malkapuram
19.	Public well	Inside the APSEB Quarters; Malkapuram
20.	Bore-well	Near H/o S. Apparao; Malkauram
21.	Bore-well	Near H. No. 61-4-27; Malkapuram
22.	Bore-well	Near H. No. 61-4-27; Malkapuram
23.	Well	Near H. No. 61-1-1A; Sriharipuram
24.	Bore-well	Opp. to Ramamandiram; North Sriharipuram
25.	Public well	Near St. Andrew's School; Sriharipuram
26.	Bore-well	Near H. No. 62-4-43; Sriharipuram
27.	Well	Inside the Elementary School; Sriharipuram
28.	Public well	Near H. No. 64-3-19; Ramnagar
29.	Well	Near H. No. 64-3-52; Ramnagar

(Contd.)

Sample No.	*Source*	*Particulars*
30.	Bore-well	Near H. No. 64-5-15; Ramnagar
31.	Well	Near Ganesh Temple; Scindhia Jn. near Shipyard
32.	Well	Adjacent to postal staff quarters; Scindhia Jn.
33.	Bore-well	Behind port quarters; Durganagar
34.	Well	Opp. to Panchadarlavari house, Mudunutula area
35.	Well	Near H. No. 63-2-162; Barma Colony
36.	Bore-well	Near Seven temples; Prakashnagar
37.	Well	Near Seven temples; Prakashnagar
38.	Bore-well	Opp. to H. No. 63-3-42; Jawaharnagar
39.	Public well	Opp. to H. No. 63-3-52; Jawaharnagar
40.	Bore-well	Near H. No. 60-27-70; Industrial colony
41.	Bore-well	Near H/o K. Simhadri; Industrial colony
42.	Bore-well	Opp. to H. No. 63-2-296; Janata colony
43.	Bore-well	H/o M.V. Apparao; Janata colony

The possibility of salt water intrusion has been ruled out by considering the following facts :

(*a*) The areas like Sindia junction, Shipyard colony, Trinadhapuram and Malkapuram, which are on the northern side of R.K. Puram are nearer to the sea. The salinity and hardness in the groundwater of these areas are very low when compared to those of R.K. Puram.

(*b*) Similarly the groundwater in Jawaharnagar and Janata Colony which are closer to the sea on the eastern side are also found to have low values of hardness and salinity compared to those of R.K. Puram.

(*c*) Another important point that should be considered is that the Visakhapatnam Port Trust draws groundwater in Malkapuram area to an extent of approximately 1.5 lakh litres a day. This water was found to be normal even after the R.K. Puram groundwater became hard and saline. If the reason for sudden increase in the hardness and salinity in R.K. Puram groundwater was due to the sea water intrusion, it would have first effected the groundwater in Malkapuram and other areas considered above.

Table 34.3 : Comparison of Hardness Values in Ramakrishnapuram Area

S. No	20-6-85	5-12-85	2-7-86	21-3-87	13-12-87
1.	1,101	967	423	388	310
2.	21,548	1,922	703	524	433
3.	27,534	7,142	5,360	2,485	1,571
4.	39,505	11,727	8,611	2,735	2,193
5.	13,168	2,363	2,204	1,527	1,015
6.	32,438	9,117	6,894	3,728	1,084
7.	825	527	476	458	316
8.	2,442	1,164	615	429	382
9.	4,788	1,270	1,142	1,112	684
10.	6,816	2,340	2,285	1,368	883
11.	6,927	2,810	2,428	1,416	792
12.	2,048	1,815	878	661	556
13.	2,124	1,929	924	821	425
14.	1,616	1,247	868	798	398
15.	982	958	703	388	304
16.	498	478	439	288	260
17.	464	494	615	3,287	3,925
18.	253	229	625	547	458
19.	350	335	763	496	296
20.	7,183	2,984	2,316	1,514	892
21.	10,883	4,321	3,619	2,086	993
22.	1,994	1,834	916	684	562
23.	497	423	566	524	364
24.	788	776	527	853	416
25.	632	615	464	709	392
26.	754	712	488	768	397
27.	658	670	482	772	406
28.	482	917	1,845	4,012	3,403
29.	589	1,345	3,696	3,352	2,996
30.	362	388	1,758	1,443	1,298
31.	368	394	439	319	302

Table 34.3 (Contd.)

S. No	*20-6-85*	*5-12-85*	*2-7-86*	*21-3-87*	*13-12-87*
32.	312	335	308	316	289
33.	156	239	352	242	198
34.	778	864	954	803	762
35.	342	378	439	284	260
36.	384	400	527	378	361
37.	416	528	591	236	218
38.	575	335	527	331	319
39.	388	442	502	307	393
40.	146	282	348	284	391
41.	218	265	527	236	491
42.	217	285	439	314	282
43.	208	278	442	306	298

Finally the industrial activities in this area are considered. Since no effluent passes through or near R.K. Puram area, the possibility of effluent invasion causing the rise in salinity is ruled out. This leaves for consideration the possibility of any single activity in that area responsible for the sudden change in the quality of water. Detailed efforts are made to collect information from various sources on this possibility.

One instance that has drawn the attention was the testing of three naphtha storage tanks each of 23.5 million litres capacity by the refinery in this area during the period of February to April, 1985. The tanks are located 15 to 20 m. from the road, that divides R.K. Puram and the refinery area. Two of these tanks are opposite to R.K. Puram and the third one is opposite to Ramnagar (Fig. 34.2).

It is general practice in coastal regions to test the new storage tanks of a refinery for leaks with sea water. It is learnt that the refinery tested the newly constructed storage tanks for leaks during February to April, 1985 using sea water.

Table 34.4 : Comparison of Chloride Values in Ramakrishnapuram Area

S. No	20-6-85	5-12-85	2-7-86	21-3-87	13-12-87
1.	235	210	208	150	128
2.	5,554	1,361	366	317	206
3.	7,290	5,100	3,670	1,856	731
4.	11,638	10,003	4,421	1,019	937
5.	3,471	2,309	1,275	462	408
6.	9,524	7,223	3,088	973	742
7.	362	314	276	262	168
8.	1,375	583	517	324	302
9.	1,469	865	799	645	317
10.	2,614	1,499	802	506	350
11.	7,182	4,196	2,570	1,893	693
12.	812	648	427	337	133
13.	665	588	552	444	262
14.	802	718	516	453	262
15.	484	452	306	183	178
16.	228	184	167	148	124
17.	190	284	383	1,174	2,268
18.	93	69	142	105	102
19.	138	138	191	174	98
20.	1,475	982	827	726	829
21.	2,423	1,964	1,204	984	516
22.	1,027	991	518	392	238
23.	174	172	186	179	122
24.	408	396	358	446	214
25.	392	385	316	439	219
26.	491	462	223	498	216
27.	318	310	215	328	208
28.	337	559	681	2,564	1,865
29.	329	689	1,442	1,382	1,080
30.	182	206	918	882	867

Table 34.4 (Contd.)

S. No	*20-6-85*	*5-12-85*	*2-7-86*	*21-3-87*	*13-12-87*
31.	193	224	246	227	168
32.	162	168	133	157	125
33.	83	112	150	12	97
34.	479	517	559	562	501
35.	157	168	207	125	109
36.	303	315	362	175	135
37.	254	384	416	122	118
38.	398	232	368	223	216
39.	150	258	332	122	143
40.	81	118	142	121	159
41.	112	146	325	174	257
42.	109	152	316	228	153
43.	92	148	321	216	165

There is no canal or channel to discharge the waste water to the sea. There is no way even to divert the used sea water into the effluent channel, which is far away from the tanks. Alternatively, the refinery could have pumped back this waste water but it requires consumption of energy. As an economic measure, the refinery authorities might have decided to discharge the sea water near the naptha storage tanks without assessing the consequences, and discharged the sea water on to the ground adjacent to tanks. Thus, there seems to be a judgement error on the part of the refinery employees which lead to the release of 470 lakh litres of the sea water on to the ground surrounding the tanks instead of releasing it into the effluent channel. This seems to be the only possible reason that could be attributed for causing this episode.

On the basis of these facts, the possibility of the seepage of the huge quantities of sea water discharged from the tanks was considered to explain the increase in salinity and hardness in the groundwater of R.K. Puram area. The contour map (Fig. 34.2) indicates the variation of hardness in the well waters. From this map, the point of discharge of sea water could be inferred and it is likely to be somewhere near the naptha storage tanks.

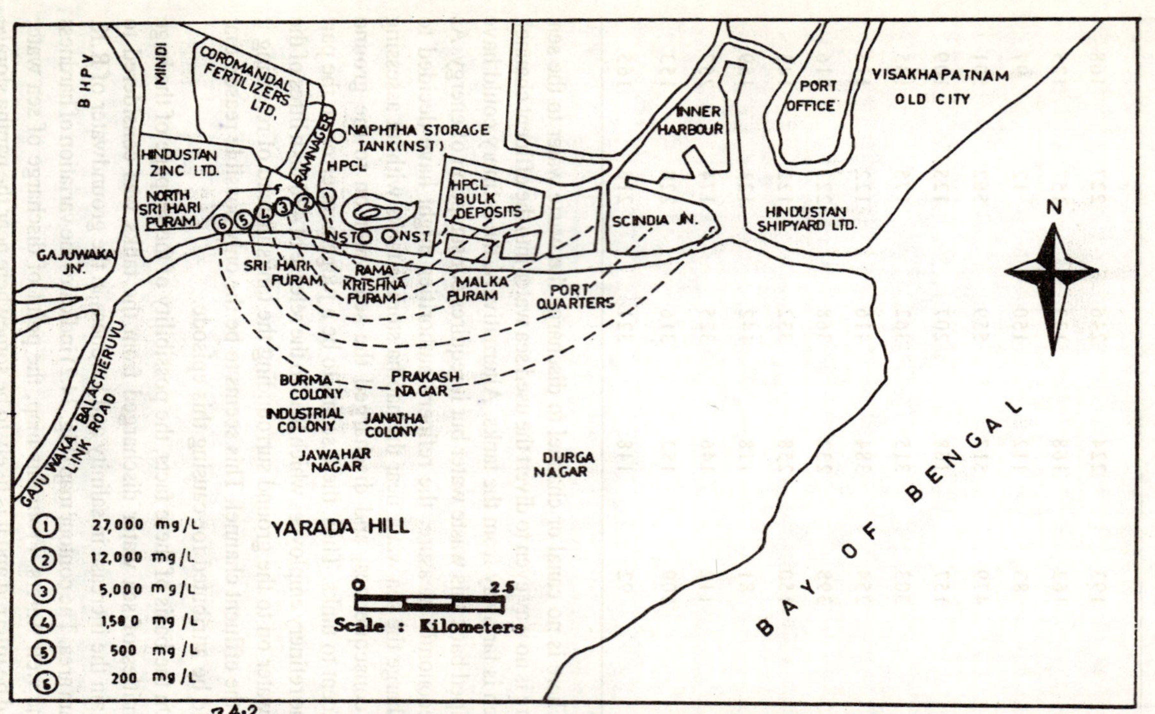

FIG. 34·2 **HARDNESS CONTOURS IN R.K. PURAM AREA OF VISAKHAPATNAM CITY**

The groundwater in Ramnagar which is opposite to the third naptha storage tank, was normal during 1985, was polluted by the sea water discharged from the third naptha storage tank during June, 1986. The magnitude of the increase in hardness and salinity in Ramnagar area is lower than that of R.K. Puram area. This can be attributed to the relatively fast movement of the discharged sea water from the new tank through the canal separating Ramnagar from the refinery and also due to the percolation from the sewage canal which will dilute the seeping salt water released from the tank. The canal carries sewage and sullage from R.K. Puram, Sriharipuram and Naval establishments to the western arm of the inner harbour and there is a continuous flow of water in the canal throughout the year.

The data given in Tables 34.3 and 34.4 further shows a decreasing trend in the hardness and salinity values of the groundwater in R.K. Puram and Ramnagar areas. This may be due to the dilution with rain water in the monsoon season. The trend of dilution is continuing and it can be predicted that the quality of water would become normal within a period of a few years with normal rainfall. Thus the contamination may not be permanent in nature. However, a slight increase in total dissolved solids, hardness and salinity in summer season is noticed. This may be due to seasonal variation, which is due to depletion of groundwater by evaporation.

From Tables 34.1 and 34.2, it can be seen that high concentrations of nitrite are present in the wells which are near the drainage canals through which the sewage and sullage flows. This may be attributed to the seepage of drainage through the soil. The concentrations of nitrate and phosphate are also high in most of the wells which may be also due to the seepage of the sewage. The concentrations of fluoride in all the wells are within the acceptable limits (W.H.O., 1984). The concentrations of sulphate in the wells where the seepage occurred is very high whereas in other wells its concentration is normal. This is also due to the infiltration of the sea water.

As explained above, the reason for the sudden increase in hardness and salinity of groundwater in R.K. Puram is attributed to the seepage of sewage discharged near the tanks. The contamination of groundwater by high chloride containing wastes or brine have been reported in four instances. Ulrich (1955) reported the field situation involving high chloride contamination of wells as Massillon, Ohio, U.S.A. Chloride content has been estimated to rise from 8 ppm to 1700 ppm in the wells. This has been attributed to infiltration of industrial wastes from an adjacent river or upward movement of deeper saline water without confirming the source. Parks (1959) reported that the concentration of groundwater by chloride from the discharge of brine from a water softening plant at Indian hill, Ohio, U.S.A. Schmidt *et al.*, (1981) also reported high chloride contamination of groundwater due to the seepage of the brine from an ice plant and water softening plant. Even though the discharge of waste water from the water softening plant was stopped, high chloride concentrations in the groundwater were observed to the present even after 26 years. Collins (1971) reviewed the contamination of groundwater by the waste water from oil and gas wells which resulted in increase of chloride concentration in few instances.

As can be seen from Table 34.1, in the case of some wells in R.K. Puram, the hardness is higher than even that of sea water. It is noteworthy that the content of the calcium is higher than that of magnesium in the well waters, whereas magnesium is in greater concentrations than calcium in sea water. To confirm the conclusion, the authors have carried out simulated studies.

The results of the simulation study are presented in Table 34.5. As can be seen from the results, the first fraction of the eluate from the column has high hardness and it slowly decreases in the subsequent fractions and attains a constant value there after. This can be explained as due to attainment of ion exchange equilibrium. Another important feature of the study is that the hardness of the eluate is increasing with the increasing concentrations of the sodium chloride solution used for leaching. The hardness of the sea water is almost doubled after passing through the soil column. Thus it is very clear that sodium is exchanging with calcium and magnesium of the soil and hence the leaches contain high hardness.

Various ions in the soil can be replaced by each other by proper choice of concentrations. Thus when sea water is rich in sodium ions passed through the soil, the sodium ions will be exchanged by the calcium ions and as a result the calcium concentration in the eluate gradually increases. The ion exchange phenomenon clearly explains the observations that in some wells the calcium is present in concentrations higher than that in sea water. The enormous increase in the concentration of calcium in well waters cannot be only due to the exchange by sodium ions. The magnesium ions present in the sea water may also exchange with calcium of the soil resulting in high calcium in the eluate. This argument appears to be resonable as the calcium content in well waters of R.K. Puram is very high compared to that of magnesium.

Table 34.5 : Hardness in the Effluent of Soil Columns Leached with Varying Concentrations of Sodium Chloride and Sea Water

Fractions	*Sodium in mg/l*													*(mg $CaCO_3$/l)*
	*0**	*50*	*100*	*150*	*200*	*300*	*500*	*1,000*	*2,000*	*3,000*	*4,000*	*5,000*	*1,0000*	*9835***
Before passing	Nil	Nil	Nil	Nil	Nil	Nil	Nil	Nil	Nil	Nil	Nil	Nil	Nil	6927
After passing :														
1.	134	147	202	257	396	835	1400	2802	3383	4202	5045	7651	12200	
2.	89	140	141	230	252	314	600	794	1468	1530	1824	2088	2611	8064
3.	71	116	116	162	203	252	352	542	972	982	1112	1220	1282	7186
4.	58	87	93	132	157	203	269	434	765	794	815	827	844	7134
5.	41	77	89	116	141	180	221	345	569	636	641	662	707	6979
6.	41	74	89	106	126	155	192	290	558	583	594	606	653	6972
7.	41	72	83	95	108	137	174	260	445	472	478	484	502	6927
8.	41	66	74	87	95	128	161	228	414	425	431	436	456	6927
9.	41	62	74	87	114	139	215	393	408	413	415	424	6927	6927
10.	41	56	74	83	86	103	132	211	277	292	297	298	316	6927
11.					99	132	180	250	265	269	271	289	6927	6927
12.					93	132	174	232	241	255	262	271	6927	6927
13.					87	132	174	232	241	255	262	271	6927	6927
14.					83	103	135	199	204	209	212	219	6927	6927
15.					81	91	128	186	191	195	198	20	6927	6927

* Distilled water ; ** Sea water.

Recommendations

(1) It is recommended to the Municipal Corporation of Visakhapatnam to supply tap water to R.K. Puram area which was implemented by them with the financial help from the refinery.

(2) It is recommended to refineries and other chemical industries to discharge the used sea water after testing the tanks, should be released in properly planned ways to avoid groundwater pollution.

Acknowledgements

The authors express their grateful thanks to Prof. M.N. Sastry, School of Chemistry, Andhra University, for his useful discussions and encouragement. The authors are also thankful to the Department of Environment, New Delhi for financial assistance in the form of a research project to the first author.

References

(APHA) Franson, M.A.H. (Ed.), American Water Works Association and Water Pollution Control Federation, (1985) Standard Methods for the Examination of Water and Wastewater, Washington D.C.

Collins, A.G., (1971) Oil and gas wells-potential polluters of the environment Jour. Water Pollution Control Fed., V. 43, pp. 2283-393.

Golterman, H.L., Glyno, R.S. and Ohnstad, M.A.M., (1978) IBP Handbook of methods for physical and chemical analysis of fresh waters. Blackwell Scientific Publications, Oxford.

Jan, T.K. and Yong, D.R., (1978) Determination of microgram amounts of some transition metals in seawater by methyl isobutyl ketone-nitric in successive extraction and flameless atomic absorption spectrophotometry. Anal. Chem. V. 50, pp. 1250.

Parks, W.W., (1959) Decontamination of groundwater at Indian oil Jour. Amer. Water Works Association, V. 51, 5, pp. 644-46.

Rao, B.P., (1971) A study in geography of port town, National Geographical Society of India, Varanasi.

Schmidt, K.D., Krancher, J.A. and Bisel Jr. G., (1981) Brine Pollution at Fresno-Twenty-six years later. Groundwater, V. 19, 1 pp. 12-19.

Ulrich, A.A., (1955) Chloride contamination of groundwater in Ohio. Jour. Amer. Water Works Association, V. 47, 5, pp. 151-52.

Vogel, A.I., (1978) Qualitative Inorganic Analysis, Longman Group Ltd., England.

W.H.O., (1984) Guidelines for drinking water quality (Vol. 2), World Health Organisation, Geneva.

35

Pollution Studies of the Swarna Rekha River in Gwalior, India

SANJAY SHARMA*,
R. MATHUR
AND
ASHA MATHUR**

Introduction

In India eighty-three per cent of the drainage basin is shared by fourteen major river systems, which account for eighty-five per cent of the surface flow and serve nearly eighty per cent of the total population in the country. Besides this there are also forty-four medium and fifty-five minor rivers which discharge nearly eighty per cent of their water during the rainy season. Rivers, which once enjoyed religious status are now reduced to open sewers. On one hand, in urban areas, where there has been a large setup of industrial units, the rivers have suffered a great setback by virtue of effluents, containing chemical constituents in undesirable concentrations, on the other hand in rural areas where the agricultural practices have switched over from conventional type to mechanized and scientific type, it has received large amounts of pesticides due to their improper use and disposal. Whether industrial pollution or agricultural pollution, whatsoever be the case, the spurt in the population is the common reason behind the pollution of rivers. Due to the flowing nature of the rivers they have been the most convenient source of disposing wastes. Voluminous information is available on the water quality and pollution status of Indian rivers (Bhaskaran *et al.*, 1965; Basu, 1966; David and Ray, 1966; George *et al.*, 1966; Saxena *et al.*, 1966; Verma and Dalela, 1975; Rama Rao *et al.*, 1978; Verma *et al.*, 1978; Mitra, 1982; Mahadevan and Krishnaswamy, 1983; Jagdale *et al.*, 1984; Raina *et al.*, 1984; Verma *et al.*, 1984; Malhotra *et al.*, 1987; Sharma *et al.*, 1989; Sharma, 1991 and Mathur *et al.*, 1991). The present study on the Swarna Rekha river is yet another study in this direction which highlights the quality of domestic wastes flowing into the channel.

Study Area and Selection of Sites

The studied area falls between latitudes 26°10′ N and 26°17′ and longitude 78°10′E. Swarna Rekha river in Gwalior is a tributory of the river Chambal. This is the solo sewage channel in the city which carries raw sewage from the heart of the city to the outskirts at Sharma Dairy Farm, where its use is made in agricultural farming. The river is stretched in the city from Hanuman Bundha to Jalalpur dam in a total stretch of 13.5 km. Historically it is said that fifty to sixty years back the water quality in this river was so pure that its water was used for drinking, bathing and other household purposes. But since then, owing to the manifold increase in the city population, status of this river has been reduced merely to an open sewage channel. Six sampling stations were established on the river, namely, at Taraganj, Jiwajiganj, Phoolbagh, Gouspura, Sharma Dairy Farm and Jalalpur. Selection of sites was based on the tendency of wastes input. The first site was established at Taraganj which has maximum input of wastes and is 1.88 km from Hanuman Bundha. This entry of sewage, however, declines on the second station which was established at Jiwajiganj, 1.52 km ahead from the station 1. The third station was established on the midway of the channel at Phoolbagh where there is no fresh sewage entry and is 2.68 km ahead the station 2. The fourth station was established at Gouspura, 3.24 km ahead from station 3. This station was second highest after Taraganj in terms of influx of sewage. Sewage is diverted directly into the agriculture lands thereby reducing the flow rate, depth and increasing the rate of sedimentation. The fifth sampling station on the sewage channel was established at Sharma Dairy Farm situated 2.28 km ahead from station 4. There is no fresh sewage entry and the conditions are relatively better. The sixth station was at Jalalpur, 1.90 km ahead from the fifth and fresh entry of sewage was nil.

Materials and Methods

The present study on the Swarna Rekha sewage channel in Gwalior has been carried out for a period of two years commencing from July 1987 to June 1989. The surveyed river is a solo domestic waste channel running through the heart of Gwalior city for a distance of 13.5 km from Hanuman Bundha to Jabalpur pickup weir. Six sampling stations were established on the sewage channel, namely, at Taraganj, Jiwajiganj, Phoolbagh, Gouspura, Sharma Dairy Farm and Jabalpur located at the distance of 1.88 km, 3.40 km, 6.08 km, 9.32 km, 11.6 km and 13.5 km respectively, from the headwaters of the river. Sewage samples were collected each month in the early hours of the day, *i.e.*, between 5.00 A.M. to 8.00 A.M. Some characteristics like, water temperature, Secchi disc transparency, depth, current velocity, dissolved oxygen and free carbon dioxide were determined on the sampling stations itself. Whereas other characteristics like turbidity, pH, electrical conductance, total alkalinity, magnesium, calcium hardness, total hardness, chloride, sulphide, sulphate, nitrite, nitrate, phosphate, silicate, sodium, potassium, chemical oxygen demand (C.O.D.), total dissolved and suspended solids (T.S., T.D.S., T.S.S.) were determined in the laboratory as per the methods prescribed in the "Standard Methods" for the examination of waters and wastewaters (APHA, 1985). The meterological data appertaining to the study period were obtained from the local meteorological laboratory at Gwalior.

Results

Detailed results of water chemistry of the Swarna Rekha are explicited in Tables 35.1-35.6. The results have been highlighted as under :

Secchi disc transparency fluctuated from 3 cm (June 1988) at station IV to 41 cm (Aug. 1988) at station VI.

Oxygenation in the sewage channel varied in a wide range fluctuating from 1.11 mg/l^{-1} (Dec. 1987) at station I to 14.20 mg/l^{-1} (Jan. 1989) at station I.

Free carbondioxide content of the sewage were very high. The variation ranged from 15.4 mg/l^{-1} (Nov. 1988) at station II to 158.4 mg/l^{-1} (Aug. 1988) at station I.

Concentration of hydrogen ions ranged from 6.68 (Jan. 1989) at station VI to 8.74 (Feb. 1988) at station III.

Total alkalinity of the sewage water samples fluctuated from 214 mg/l^{-1} (Aug. 1988) at station I to 624 $mg.l^{-1}$ (May 1989) at station IV.

Total hardness of the sewage waters varied from 138 $mg.l^{-1}$ (June and Aug. 1988) at station I to 522 $mg.l^{-1}$ (May 1989) at station IV.

Chloride content of the sewage ranged from 75 $mg.l^{-1}$ (Dec. 1987) at station III to 197 $mg.l^{-1}$ (May 1989) at station IV.

Variation in the sulphide values of the sewage waters ranged from nil at different stations and months during the year July 1988-June 1989 to 3.825 $mg.l^{-1}$ (Aug. 1988) at station IV.

Sulphate content of the sewage were nil (Dec. 1988 and Aug. 1988) at station II and III whereas their maximum concentration (84 $mg.l^{-1}$) were recorded in May 1988 at station V.

Values of nitrite were recorded to be nil in the month of Oct. 1988 throughout the river. Its maximum concentration (0.560 $mg.l^{-1}$) was recorded in August 1988 at station II.

Table 35.1 : Monthly variations in physico-chemical characteristics at station I of the Swarna Rekha sewage channel, Gwalior during July 1987–June 1989.

Months	*Transpa-rency cm*	*D.O. mg/l[-1]*	*F.CO$_2$ mg/l[-1]*	*pH units*	*Total alkali-nity*	*Total hardness*	*Chlor-ide*	*Sulp-fide*	*Sulph-ate*	*Nitr-ite*	*Nitr-ate*	*Phosp-hate*	*Sili-cate*	*Sod-ium*	*Potass-ium*	*C.O.D.*	*T.D.S*
					HCO$_3$ as CaCO$_3$												
July 1987	05	4.56	46.2	7.20	564	280	145	0.425	11.0	0.490	1.683	1.04	52.0	6.20	4.65	452	1000
August	10	3.28	41.8	7.50	324	206	85	0.850	18.0	0.400	1.285	0.96	56.0	6.88	5.35	408	660
September	07	2.00	30.8	7.62	328	200	85	0.425	24.0	0.250	0.797	1.04	37.0	8.63	7.00	496	520
October	06	11.14	37.4	7.40	330	230	127	0.425	18.0	0.180	0.443	0.76	16.0	8,00	6.65	376	440
November	08	4.14	41.8	7,62	360	180	140	0.425	29.0	0.205	0.487	0.88	10.0	8.20	6.35	420	420
December	07	1.11	39.6	7.88	302	206	82	0.850	35.0	0.248	0.443	1.04	32.0	7.35	6.08	304	580
January 1988	09	5.96	74.8	7.75	438	266	160	0.850	42.0	0.032	0.665	1.20	85.5	6.70	4.00	396	600
February	08	8.45	114.4	8.65	416	236	143	1.275	47.5	0.028	0.554	1.25	72.0	6.20	4.95	396	540
March	10	2.28	149.6	7.60	358	204	152	1,275	38.0	0.032	0.797	1.20	84.5	7.30	5.63	244	360
April	08	1.58	134.2	7.67	402	306	152	1.700	33.0	0.031	0.930	1.35	60.0	6.50	5.65	312	400
May	06	1.18	145.2	7.80	480	372	173	1.700	47.5	0.034	0.842	1.39	170.0	8.20	6.38	380	560
June	04	5.80	24.2	8.03	240	138	99	1.275	46.0	0.038	0.642	1.22	188.0	5.50	6.08	212	320
July	05	7.65	33.0	7.33	282	158	125	0.850	39.0	0.035	0.509	1.22	152.0	3.00	5.35	172	1020
August	05	1.30	158.4	7.12	214	138	111	2.550	16.0	0.440	1.085	0.86	45.0	2.63	6.34	264	440
September	17	13.34	61.6	7.12	242	154	88	0.425	18.5	0.119	0.443	1.20	30.0	1.88	8.28	188	220

October	05	5.91	19.8	7.25	224	170	92	0.425	16.5	nil	0.354	0.46	18.0	8.85	7.35	136	360
November	12	7.07	22.0	7.40	274	200	100	nil	35.0	0.46	0.443	0.95	32.0	8.70	6.70	160	340
December	06	6.10	55.0	7.25	346	242	160	1.700	16.0	0.032	0.443	1.23	52.5	5.40	6.08	196	440
January 1989	05	14.20	39.6	7.10	236	200	90	0.425	15.0	0.044	nil	1.01	99.0	3.60	3.05	192	240
February	06	11.42	28.6	7.62	292	226	132	0.425	17.0	0.039	0.421	1.20	98.0	4.88	5.35	252	480
March	09	7.60	39.6	7.72	324	230	147	0.850	33.0	0.041	0.554	1.25	125.0	6.93	5.68	240	280
April	08	6.32	46.2	7.81	362	274	158	0.850	36.0	0.041	0.842	1.30	140.0	8.20	4.98	352	360
May	05	3.30	46.2	7.88	552	410	186	1.275	42.5	0.041	0.687	1.45	180.0	8.05	6.08	388	480
June	07	4.43	37.4	7.96	392	250	172	1.275	36.0	0.041	0.886	1.25	155.0	6.05	4.94	284	600

Table 35.2 : Monthly variations in physico-chemical characteristics at station II of the Swarna Rekha sewage channel, Gwalior during July 1987–June 1989.

Months	*Transparency cm*	*D.O. mg/l⁻¹*	*F.CO₂ mg/l⁻¹*	*pH units*	*Total alkalinity HCO₃*	*Total hardness as CaCO₃*	*Chloride*	*Sulpfide*	*Sulphate*	*Nitrite*	*Nitrate*	*Phosphate*	*Silicate*	*Sodium*	*Potassium*	*C.O.D.*	*T.D.S*
July 1987	07	3.32	26.4	7.14	560	308	148	0.850	29.5	0.450	1.772	1.05	59.0	6.75	4.65	548	960
August	19	4.11	35.2	7.26	306	218	96	0.850	34.0	0.390	1.506	1.00	63.0	8.25	5.00	364	560
September	11	4.80	28.6	7.38	260	192	91	0.425	36.0	0.300	1.152	1.10	32.0	10.25	6.65	480	920
October	13	6.91	37.4	7.66	308	226	131	0.455	35.0	0.200	0.709	1.12	24.0	8.70	6.68	428	360
November	16	6.02	33.0	7.68	346	194	140	0.850	33.0	0.210	0.554	1.20	9.0	7.90	6.70	452	760
December	19.5	4.43	37.4	7.82	394	204	95	0.850	23.0	0.260	0.598	1.22	24.0	8.24	6.38	376	880
January 1988	20	7.08	59.4	7.81	486	.284	151	1.275	29.0	0.030	0.687	1.35	91.0	7.45	5.00	364	640
February	18	8.17	90.2	8.60	500	254	148	1.275	43.5	0.031	0.753	1.45	110.0	6.35	6.03	412	600
March	20	3.43	123.2	7.82	462	202	150	1.700	40.0	0.030	0.753	1.25	100.0	7.45	6.35	368	480
April	15	2.78	127.6	7.70	386	292	151	1.700	36.0	0.031	1.019	1.35	81.0	6.70	6.35	344	400
May	11	1.96	125.4	7.78	426	402	177	2.125	59.0	0.038	0.886	1.40	190.0	8.25	6.65	412	640
June	08	6.19	19.8	8.22	286	146	101	0.850	58.0	0.034	0.465	1.30	158.0	5.25	5.35	272	520
July	11	7.65	26.4	7.22	346	214	136	0.850	43.5	0.039	0.532	1.20	154.0	4.50	6.33	268	920
August	04	1.30	140.8	7.18	284	144	137	0.425	14.5	0.560	1.019	1.01	46.0	3.13	6.05	384	360
September	13	10.76	66.0	7.18	238	162	86	0.850	14.5	0.025	0.399	1.10	10.0	8.85	8.33	224	220

October	08	4.22	30.8	7.31	352	200	125	0.425	13.5	nil	nil	0.80	6.0	8.65	8.04	152	480
November	07	8.86	15.4	7.47	218	178	130	0.425	60.0	0.239	0.886	0.47	101.0	8.80	7.38	172	260
December	05	7.29	72.6	7.35	270	248	160	1.275	nil	0.034	0.443	0.94	24.5	7.50	6.10	240	580
January1989	08	12.53	39.6	7.15	276	196	110	1.275	16.5	0.034	nil	1.04	176.0	2.88	5.00	72	260
February	14	9.38	28.6	7.60	308	268	141	nil	18.5	0.045	0.465	1.35	290.0	4.90	5.60	288	600
March	16	7.60	55.0	7.86	368	250	151	0.850	37.0	0.043	0.598	1.35	160.0	7.00	7.00	366	360
April	14,5	4.35	44.0	7.92	446	304	170	1.275	40.0	0.042	0.842	1.40	165.0	8.58	6.03	352	440
May	11	2.48	37.4	9.96	608	454	191	2.125	55.5	0.050	1.218	1.45	195.0	8.38	7.04	416	560
June	13	4.83	30.8	8.11	426	286	183	1.275	43.5	0.041	0.997	4.30	175.0	6.75	5.30	364	760

Table 35.3 : Monthly variations in physico-chemical characteristics at station III of the Swarna Rekha sewage channel, Gwalior during July 1987–June 1989.

Months	Transparency cm	D.O. mg/l^{-1}	FCO_2 mg/l^{-1}	pH units	Total alkalinity HCO_3	Total hardness as $CaCO_3$	Chloride	Sulphide	Sulphate	Nitrite	Nitrate	Phosphate	Silicate	Sodium	Potassium	C.O.D.	T.D.S
July 1987	09	6.22	33.0	7.60	330	190	86	0.425	1.0	0.338	1.174	0.69	20.0	4.00	4.00	352	480
August	16	2.88	68.2	7.42	370	276	102	0.425	16.0	0.296	1.351	0.75	38.0	6.20	4,34	444	600
September	11	2.40	24.2	7.63	366	260	95	0.425	26.0	0.238	0.554	0.94	20.0	8.75	6.34	448	720
October	9.5	5.76	48.4	7.24	336	280	132	0.850	42.5	0.129	0.532	0.87	10.0	7.13	5.35	380	520
November	14	3.01	55.0	7.68	314	276	141	0.425	38.0	0.180	0.465	0.97	1.6	7.25	7.38	488	320
December	12	3.67	33.0	7.70	290	200	75	0.425	20.2	0.200	0.443	1.05	17.0	6.20	6.10	200	420
January 1988	11	2.24	46.2	7.56	298	256	134	0.850	39.0	0.033	0.554	0.96	89.0	6.15	4.65	224	320
Febuary	11	3.34	107.8	8.74	364	224	141	1.275	47.5	0.030	0.642	1.20	84.5	5.70	5.61	240	400
March	14	12.19	156.2	7.75	352	196	140	1.700	46.0	0.025	0.665	1.15	88.0	7.00	6.32	208	420
April	15	3.56	138.6	7.73	428	252	148	1.700	31.0	0.030	0.842	1.25	71.5	5.45	6.05	272	560
May	10	2.74	127.6	7.86	416	452	169	1.700	56.0	0.032	0.753	1.35	179.0	8.05	6.00	464	780
June	06	4.64	30.8	8.06	284	162	114	0.850	60.0	0.037	0.433	1.20	150.0	4.00	6.04	168	560
July	07	5.74	28.6	7.31	268	154	131	0.425	55.0	0.033	0.399	1.15	67.0	3.05	6.10	160	820
August	09	9.54	81.6	728	314	202	120	2.125	nil	0.238	nil	1.15	61.0	2.88	6.98	284	620
September	14	8.60	33.0	7.37	304	202	96	0.850	23.0	0.017	0.642	1.20	nil	6.28	6.34	368	300

October	13	5.91	26.4	7.40	300	296	112	0.425	15.0	nil	nil	0.65	8.0	8.78	5.35	120	280
November	05	7.49	26.4	7.46	276	220	112	0.425	36.5	0.189	1.330	0.63	81.0	8.31	6.65	368	420
December	08	4.45	55.0	7.37	382	420	125	1.700	22.0	0.028	nil	1.30	31.2	9.50	6.04	148	740
January 1989	09	10.44	44.0	7.12	292	230	103	1.275	17.5	0.029	0.753	1.04	171.0	7.45	3.00	168	400
February	11	11.83	33.0	7.66	352	254	136	0.850	16.5	0.035	0.772	1.25	270.0	1.95	3.70	272	560
March	13	6.40	41.8	7.60	386	262	160	0.425	34.0	0.039	0.465	1.20	140.0	6.70	6.00	380	320
April	11	6.32	39.6	7.98	512	328	183	0.425	34.5	0.037	0.731	1.30	160.0	7.52	6.01	484	540
May	09	4.13	28.6	8.24	576	490	189	1.700	45.5	0.041	1.130	1.45	195.0	7.20	6.34	412	520
June	11	5.64	35.2	8.18	484	372	194	0.850	32.0	0.033	0.797	1.15	155.0	5.88	4.35	372	600

Table 35.4 : Monthly variations in physico-chemical characteristics at station IV of the Swarna Rekha sewage channel, Gwalior during July 1987–June 1989.

Months	*Transparency cm*	*D.O. mg/l^{-1}*	*F.CO_2 mg/l^{-1}*	*pH units*	*Total alkalinity HCO_3*	*Total hardness as $CaCO_3$*	*Chloride*	*Sulpfide*	*Sulphate*	*Nitrite*	*Nitrate*	*Phosphate*	*Silicate*	*Sodium*	*Potassium*	*C.O.D.*	*T.D.S*
July 1987	20	4.97	30.8	7.23	426	200	109	0.850	37.5	0.490	2.038	1.05	89.0	7.00	5.35	200	1060
August	11	2.88	30.8	7.31	462	304	125	0.425	24.8	0.430	1.794	1.05	75.0	7.50	5.05	328	620
September	13	6.80	26.4	7.93	470	300	105	1.275	54.5	0.320	1.351	1.19	70.0	10.50	7.35	180	1540
October	12	8.45	52.8	7.54	344	296	103	1.275	40.7	0.230	0.842	1.08	50.0	8.25	8.0	296	840
November	18	4.14	96.8	7.42	494	304	130	1.275	44.5	0.250	0.731	1.30	105.0	8.85	8.30	260	680
December	14	7.76	103.4	7.69	288	268	112	1.700	25.0	0.210	0.775	1.35	76.0	11.75	7.65	244	320
January 1988	17	5.96	63.8	7.61	378	328	136	1.700	45.3	0.042	0.620	1.25	130.0	8.75	6.35	204	520
February	09	4.45	112.2	8.68	382	244	140	1.275	58.5	0.048	0.952	1,43	147.0	9.30	6.70	220	440
March	08	6.85	112.2	7.99	452	254	154	1.700	52.5	0.041	1.041	1.40	134.0	9.25	7.08	256	700
April	08	3.96	107.8	7.83	480	316	163	2.125	58.8	0.061	1.174	1.46	142.0	9.60	7.04	240	700
May	05	3.14	94.6	7.93	572	446	174	2.975	70.5	0.052	1.351	1.55	210.0	10.25	8.65	576	920
June	03	2.71	44.0	7.78	434	260	174	0.850	47.8	0.037	0.817	1.25	170.6	5.25	6.36	152	1060
July	06	6.88	19.8	7.46	496	212	152	2.975	46..5	0.041	0744	1.16	162.0	4.38	7.05	340	680
August	06	11.27	116.8	7.43	482	316	187	3.825	11.0	0.344	0.819	1.52	83.0	3.53	6.30	148	940
September	13	8.60	37.4	7.75	428	300	135	0.850	24.8	0.027	1.794	1.35	11.0	9.50	6.08	76	560

October	14	5.91	39.6	7.43	428	344	153	0.850	10.5	nil	1,449	1.03	nil	7.90	6.63	64	420
November	08	7.90	26.4	7.40	380	290	155	1.700	52.3	0.202	2.538	1.12	117.0	8.85	7.32	136	420
December	05	4.05	59.4	7.44	488	504	180	0.850	26.5	0.043	0.753	1.27	60.5	17.50	6.10	124	560
January 1989	09	12.53	44.0	7.16	364	350	133	nil	18.0	0.084	nil	0.83	158.0	8.40	4.00	120	520
February	10	10.20	39.6	7.79	392	344	144	0.850	19.5	0.050	0.687	1.40	330.0	4.30	5.05	264	800
March	12	9.20	37.4	7.92	434	342	170	1.275	34.5	0.054	0.952	1.45	181.0	7.20	6.68	372	580
April	10	7.51	46.2	8.03	536	350	179	2.125	37.5	0.058	1.108	1.52	193.0	9.88	8.35	416	620
May	08	6.60	35.2	8.28	624	522	197	3.400	58.4	0.058	1.263	1.60	210.0	9.65	9.00	548	800
June	09	3.63	39.6	8.13	510	308	180	1.700	45.5	0.051	0.908	1.30	154.0	7.08	7.03	400	800

Table 35.5 : Monthly variations in physico-chemical characteristics at station V of the Swarna Rekha sewage channel, Gwalior during July 1987–June 1989.

Months	*Transparency cm*	*D.O. mg/l⁻¹*	*F.CO₂ mg/l⁻¹*	*pH units*	*Total alkalinity HCO₃*	*Total hardness as CaCO₃*	*Chloride*	*Sulpfide*	*Sulphate*	*Nitrite*	*Nitrate*	*Phosphate*	*Silicate*	*Sodium*	*Potassium*	*C.O.D.*	*T.D.S*
July 1987	16	6.22	41.8	7.64	492	220	98	0.425	24.2	0.420	1.883	1.04	57.8	6.25	4.30	180	940
August	11	4.11	35.2	7.52	356	282	106	0.425	11.0	0.300	0.731	0.80	39.0	4.20	3.30	44	46-
September	11	4.80	30.8	7.87	380	266	101	0.425	30.0	0.286	0.997	1.04	52.6	7.50	5.35	212	420
October	18	7.29	48.4	7.63	426	314	149	0.425	23.2	0.268	0.886	1.00	67.0	7.15	4.96	104	1160
November	21	5.64	44.0	7.53	472	282	153	1.275	15.0	0.302	1.254	1.07	91.0	7.35	6.00	84	460
December	26	6.65	72.6	7.67	328	268	120	0.850	27.0	0.266	0.987	1.22	59.5	9.25	7.00	100	380
January 1988	20	8.20	46.2	7.41	416	280	130	1.275	38.5	0.037	0.810	1.30	90.0	8.00	5.68	188	300
February	26	7.05	103.4	8.42	472	270	129	1.275	54.5	0.043	0.784	1.34	150.0	8.65	6.63	196	500
March	19	8.38	94.6	8.18	460	228	140	1.700	63.0	0.039	0.908	1.35	145.0	8.80	7.00	228	540
April	17	5.93	92.4	7.90	472	300	156	1.700	53.0	0.043	1.041	1.45	140.0	8.95	6.70	244	640
May	08	3.92	96.8	7.97	500	424	171	2.550	84.0	0.047	1.108	1.52	183.0	9.75	7.60	480	840
June	10	4.26	39.6	7.88	384	248	168	2.125	37.0	0.033	0.952	1.14	137.0	5.00	5.63	124	1100
July	08	7.27	22.0	7.53	352	230	147	2.550	41.0	0.038	0.842	1.12	151.0	4.25	6.68	316	600
August	16	13.87	39.6	7.32	472	308	168	1.700	47.5	0.201	0.598	1.56	50.6	3.13	6.08	92	600
September	12	3.44	28.2	7.32	444	328	163	0.425	37.5	0.035	0.567	1.05	20.0	10.16	6.01	76	920

October	20	10.14	46.2	7.08	450	348	160	0.850	9.5	nil	1.883	1.01	57.0	7.00	5.35	48	720
November	22	6.66	39.6	7.12	456	290	164	1.700	40.3	0.220	3.336	1.04	103.0	7.55	6.05	100	480
December	20	5.67	85.8	7.45	400	368	184	0.850	17.5	0.053	0.846	0.86	69.3	15.05	6.06	44	700
January 1989	11	10.86	55.0	6.96	398	408	135	0.425	17.3	0.042	nil	1.01	158.4	8.95	4.65	228	360
February	13	11.42	35.2	7.71	376	358	141	nil	18.5	0.045	0.599	1.28	284.0	4.05	4.35	280	780
March	17	8.40	44.0	7.89	420	304	146	0.850	45.5	0.051	0.886	1.48	173.0	7.05	5.61	324	520
April	14	7.90	52.8	7.98	486	326	168	1.700	44.3	0.056	0.952	1.48	177.0	8.65	7.06	344	660
May	10	7.02	46.2	8.15	588	478	183	2.975	58.0	0.057	1.152	1.60	188.0	9.42	8.28	448	700
June	13	5.24	33.0	8.04	452	330	172	2.125	46.0	0.042	0.797	1.34	126.0	7.25	6.74	324	860

Table 35.6 : Monthly variations in physico-chemical characteristics at station VI of the Swarna Rekha sewage channel, Gwalior during July 1987–June 1989.

Months	*Transparency cm*	*D.O. mg/l⁻¹*	*F.CO₂ mg/l⁻¹*	*pH units*	*Total alkalinity HCO_3*	*Total hardness as $CaCO_3$*	*Chloride*	*Sulpfide*	*Sulphate*	*Nitrite*	*Nitrate*	*Phosphate*	*Silicate*	*Sodium*	*Potassium*	*C.O.D.*	*T.D.S*
July 1987	26	8.29	50.6	7.50	500	212	92	0.425	20.3	0.400	1.347	1.00	61.0	5.25	4.00	120	720
August	19	7.40	48.4	7.48	360	238	101	0.425	8.3	0.270	0.554	0.76	28.0	3.88	3.00	40	520
September	23	5.60	33.0	7.62	372	250	982	0.425	26.3	0.280	0.687	1.00	34.0	7.13	3.35	124	600
October	24	6.14	52.8	7.40	390	268	134	0.425	20.0	0.200	0.576	0.98	60.8	6.60	4.05	72	580
November	30	4.89	52.8	7.40	420	272	141	0.850	32.0	0.350	2.082	1.10	83.0	7.25	5.35	36	400
December	34	7.02	66.0	7.53	336	266	138	0.425	24.6	0.300	0.997	1.16	55.5	9.75	5.65	52	460
January 1988	40	9.32	52.8	7.26	390	294	129	0.850	33.0	0.036	0.465	1.05	78.5	6.25	4.65	108	400
February	33	6.31	68.2	8.40	430	246	124	0.850	58.2	0.040	0.620	1.36	145.0	8.20	6.00	128	480
March	30	9.14	79.2	8.11	502	212	138	1.275	54.5	0.041	1.028	1.31	135.0	8.75	6.65	192	720
April	26	6.72	77.0	7.96	428	280	158	1.700	59.6	0.044	0.886	1.46	100.0	8.80	6.00	236	880
May	20	5.10	74.8	8.02	496	394	170	2.550	72.5	0.049	1.019	1.51	165.0	9.30	7.01	352	720
June	24	4.64	46.2	7.96	380	242	163	1.700	50.4	0.028	0.886	1.08	123.0	6.20	6.00	152	760
July	30	8.79	24.2	7.36	360	222	140	1.700	33.8	0.038	0.817	1.10	155.0	3.88	6.35	256	580
August	41	11.27	35.25	7.28	450	294	153	2.125	43.5	0.241	0.744	1.50	57.0	3.38	5.35	72	600
September	22	2.58	22.0	7.30	454	254	155	1.275	38.9	0.039	0.656	1.24	40.8	7.50	-	104	900

October	32	6.34	48.4	7.15	350	238	128	nil	12.0	nil	2.215	0.96	83.0	6.50	3.70	56	560
November	38	8.74	30.8	7.21	364	256	135	1.275	44.2	0.234	3.336	1.40	107.0	7.95	5.67	08	540
December	20	6.48	70.4	7.49	468	450	167	nil	23.0	0.150	1.294	1.14	74.0	14.50	6.00	60	600
January 1989	15	11.70	68.2	6.88	376	424	128	0.850	16.9	0.038	nil	1.12	137.6	5.40	5.05	32	440
February	16	11.83	41.8	7.45	364	348	138	nil	18.0	0.048	0.554	1.28	260.0	3.95	5.35	212	540
March	21	7.20	50.6	7.93	412	282	140	0.425	38.5	0.050	0.908	1.40	160.0	6.20	5.30	312	540
April	20	9.48	55.0	8.02	480	334	164	1.275	44.5	0.057	0.839	1.44	172.0	8.90	6.02	324	600
May	15	7.43	41.8	8.19	574	434	176	2.125	56.3	0.058	1.063	1.53	180.0	8.35	7.09	376	1020
June	17	6.05	30.8	7.98	462	298	167	1.275	43.3	0.043	0.673	1.40	97.0	7.00	5.60	256	760

Nitrate concentration in the sewage ranged from nil (different months and stations of the year July 1988-June 1989 to 3.336 $mg.l^{-1}$ (Nov. 1988) at station V and VI.

Concentration of phosphates were low and ranged between 0.46 $mg.l^{-1}$ (Oct. 1988) at station I to 1.60 $mg.l^{-1}$ (May 1989) at stations IV and V.

Sodium content of the sewage ranged from 1.88 $mg.l^{-1}$ (Sept. 1988) at station I to 17.5 $mg.l^{-1}$ (Dec. 1988) at station IV.

Potassium content of the sewage ranged from nil (Sept. 1988) at station VI to 9.0 $mg.l^{-1}$ (May 1989) at station IV.

Fluctuations in the chemical oxygen demand values were wide and ranged between 8 $mg.l^{-1}$ (Nov. 1988) at station VI to 576 $mg.l^{-1}$ (May 1988) at station IV.

Variation in the total dissolved solids values varied from 220 $mg.l^{-1}$ (Sept. 1988) at station I and II to 1540 $mg.l^{-1}$ (Sept 1987) at station IV.

Discussion

In the Swarna Rekha sewage channel the water temperature ranged from 7° to 32°C. Transparency depends upon the turbidity of the water. Rise in the values of turbidity results in fall in transparency. Throughout the study period values of Secchi disc transparency fluctuated between 3 cm to 41 cm. In riverine systems, generally minimum transparency is recorded in rainy season due to run-off waters (Dutta and Malhotra, 1986; Sharma, 1988) and its maximum values are noted in winter months (Vass *et al.*, 1977). Unlike this an opposite phenomena of rainy season maxima and summer minima in transparency values was observed in the Swarna Rekha sewage channel. This is due to the fact that in rivers the "already clear water" receives turbid run-off waters during monsoon, thereby decreasing the transparency. On the otherhand, in sewer systems, this run-off waters not only dilute the sewage but also wash off the accumulated filth on the river bed, thus increasing the values of transparency. Water scarcity in summer months lead to discharge of relatively more concentrated sewage which thereby lowers the transparency.

Depth, which varies with the water level fluctuated widely from station to station and from season to season. On all the stations, the sewage depth was maximum in rainy season. At station IV, use of raw sewage in agriculture farming lowers the water current in the river, increasing the rate of sedimentation which resulted in low depth at this station. Velocity of the current is swiftest at the centre and near the surface as frictional forces at the banks and bottom decrease the flow. Flowing waters can be categorised as erosional, depositional and intermediate type. Erosional areas are dominated by rapid flow and removal of finer sediments, while depositional areas are characterised by slow currents and the deposits of the finer sediments (Shelford, 1914; Shelford and Eddy, 1929; Moon, 1939). The current velocity of the sewage ranged from 6 cm. sec. to 43.5 cm. sec. and thus it can be categorized under the river of intermediate category, in which the water flow is neither rapid nor slow. Throughout the year, water flow remains more or less the same except during the periods of flood in rainy season.

Except at station III, on all other stations throughout the study period, the minimum standard deviation in D.O. values was noted in summer months, however, the seasonal mean was minimun throughout the river during summer months which indicates that during summer period the D.O. content remains more or less stable and low as compared to other seasons. Solubility of oxygen is greatly affected by surface agitation (Mercer, 1967). Low water levels and rough bed at station I create surface agitation which results in increased D.O. despite higher pollution load. Thus the determining factors for oxygen in water are low water level and roughness of the river bed.

Free carbon dioxide which accumulates due to microbal activity, respiration by organism and atmospheric diffusion, acidify and water by the formation of the carbonic acid. In the present study the values of free carbon dioxide with the maximum standard deviation were noted in summer months which is in agreement with previous studies made on other lotic systems (Sharma *et al.*, 1989; Sharma, 1991).

Values of Chemical oxygen demand in the present study ranged from 8 mg/l to 576 mg/l depicting no definite pattern in seasonal variations. Turbidity values in the present study were high ranging from 24 J.T.U. to 32 J.T.U.

pH values ranging between 6.5 to 8.5 are considered as favourable for fish culture and aquatic life propagation (ISI, 1982). Throughout the study period pH at different stations of the sewage channel fluctuated from 6.88 to 8.74 units, hence it was found to be suitable for fish culture.

Values of conductance in the sewage channel fluctuated from 462 μ mhos/cm. to 1716μ mhos/cm. in winter and summer months respectively which is in agreement with the study of Mathur *et al.*, (1991) who reported minimum conductance in winter months and Upadhyaya and Roy (1982) who reported maximum conductance in summer months.

Values of total alkalinity in present study ranged from 214 mg/l^{-1} to 624 mg/l^{-1} which are much higher than those reported by David (1963), Vass *et al.*, (1977) and Raina *et al.*, (1984). Alkalinity recorded, was solely due to the presence of bicarbonates. Maximum alkalinity was recorded in summer months which is in agreement with the study of Sangu and Sharma (1985).

Magnesium content in the present study fluctuated from 12.15 mg/l^{-1} to 89.91 mg/l^{-1}. The minimum values of magnesium were noted in rainy season. In the present study calcium hardness ranged from 46 mg/l^{-1} to 256 mg/l^{-1}. Maximum values of calcium hardness were recorded in summer season.

The total hardness in the Swarna Rekha ranged between 138 mg/l^{-1} to 522mg/l^{-1}. On the basis of classifications given by Sawyer and McCarty (1967), waters of this river can be categorised as very hard.

Concentration of the chloride in the present was found ranging from 75 mg/l^{-1} to 197mg/l^{-1} which is much higher than the studies of Upadhyaya and Roy (1982) and Chattopadhya *et al.*, (1984). Lowest and highest concentration of chloride was recorded in winter months respectively.

In the present study, sulphate levels in the sewage fluctuated from 0 to 84 mg/l^{-1} which is slightly higher than those reported by Paterson and Nursall (1975) in north Saskatchewan river at Edmonton.

In the present study values of nitrites ranged from 0 to 0.56 mg/l^{-1} which is higher than those reported by Sengupta *et al.*, (1988). Nitrite contents in the sewage fluctuated from 0 to 336 mg/l^{-1} which is higher than those reported by Saxena *et al.*, (1966).

High concentrations of phosphate is indicative of organic pollution. The values of phosphate fluctuated from 0.46 mg/l^{-1} to 1.60 mg/l^{-1} which are higher than those reported by Mahadevan and Krishnaswamy (1983) and Sangu and Sharma (1985). Seasonal mean values of phosphates were minimum in rainy season and maximum in summer season throughout the sewage channel.

Silicates in sewage waters ranged from nil to 330 mg/l^{-1}. Seasonal mean of silicate contents throughout the sewage channel were minimum in rainy season and maximum in summer season.

Domestic sewage is an important source of sodium to the fresh waters. In the present study, its values in sewage fluctuated from 1.88 to 17.5 mg/l^{-1} which are much lower than the values reported by Mitra (1982) and Upadhyaya and Roy (1982).

Potassium which ranks seventh among the elements in order of abundance, in the present study was found at the level of nil to 9.0 mg/l^{-1} which is lower than those reported by Mitra (1982) and Upadhyaya and Roy (1982).

Total solids which measure the amount of all kinds of solids in water, fluctuated from 380 to 3400 mg/l^{-1}. Total dissolved solids which denote mainly the various kinds of minerals present in water and do not contain any gas or colloid fluctuated from 220 to 1540 mg/l^{-1}. Total suspended solids which denote the suspended impurities present in the water and in most cases are organic in nature, in the present study fluctuated from 20 to 2440mg/l^{-1}.

Thus in the above study it was observed that meagre sanitation facilities accompanied with short water supply are the reasons which had deteriorated the status of the Swarna Rekha river into a sewage channel. The town sewage which is freely released into the river round the year has caused heavy siltation in the waterway. This along with the malpractice of dumping garbage from the residential areas into the river have collectively reduced the carrying capacity of the river. Values of the major parameters like conductance, total alkalinity, total hardness, chloride, sulphide, sulphate, nitrate, phosphate, silicate C.O.D., sodium and solids, etc., fluctuated widely from station to station which is an indication of pollution from anthropogenic origin. Thus, it can be concluded that the deteriorating river water quality can be checked by providing effective sanitation facilities in the areas of dense habitation on the banks of the river. Dredging of the sewage channel should also be done at regular intervals which will enhance the sewage flow.

References

APHA, (1985) Standard methods for examination of water and waste water. American Public Health Association, AWWA, WPCF, New York, 16th Edn.

Basu, A.K., (1966) Studies on the effluents pulp and paper mills and its impact in bringing physicochemical changes around the several discharge points in the Hooghly river estuary. J. Instn. Engrs. India, V. 48(10), pp. 108-16.

Bhaskaran, T.R. Chakrabarty, R.N. and Trivedy, R.C., (1965) Studies on the river pollution-I. Pollution and self purification of Gomti river near Lucknow. J. Instn. Engrs. India. V. 15(6), pp. 39-50.

Chattopadhya, S.N. Routh, T. Sarma, V.P. Arora, H.C. and Gupta, R.K. (1984) A short term study on the pollutional status of the river Ganga in Kanpur Region. Ind. J. Environ. Hlth. 26(3), pp. 244-55.

David, A., (1963) Report on fisheries survey of the river Gandak (North Bihar). Sur. Rep. Cent. Inl. Fish. Res. Inst. Barrackpore, (1) 24, pp. (Mimeo).

David, A. and Ray, A.P., (1966) Studies on the pollution of the river Daha (Bihar) by sugar and distillery wastes. Environ Hlth. 8, pp. 6-35.

Dutta, S.P.S. and Malhotra, Y.R., (1986) Seasonal variations in the macrobenthic fauna of Gadigarh stream (Miran Sahib), Jammu, Indian J. Ecol. 13(1), pp. 138-45.

George, M.G., Qasim, S.Z. and Siddique, A.O., (1966) A limnological survey of the river Kali with special reference to fish mortality. Environ. Hlth. 8(4), pp. 262-69.

ISI, (1982) Indian Standards tolerance limits for inland surface waters subject to pollution. Indian Standards Institution, (IInd revision) IS : 2296-1982.

Jagdale, M.H., Salunkhe, M.M., Vibhute, Y.B., Khodaskar, S.B. and Ghuge, K.N., (1984) Pollution of Godavari river water at Nanded. Poll. Res. 3(2), pp. 83-84.

Mahadevan, A. and Krishnaswamy, S., (1983) A quality profile of river Vaigai (S. India). Ind. J. Environ. Hlth 25(4), pp. 288-99.

Malhotra, Y.R., Dutta, S.P.S. and Singh, R., (1987) Abiotic limnology of Neeru Nullah, Bhadarah. In : Perspectives in Hydrobiology (K.S. Rao & S. Shrivastava Ed.). Vikram University, Ujjain, India, Sec. IV (31), pp. 161-63.

Mathur, R., Sharma, R.K., Nand, K.C. and Sharma, S., (1991) Water quality assessment of the river Chambal over the stretch of National Chambal Sanctuary in Madhya Pradesh. Indian J. Ecd. 18(1) pp.1-4.

Mercer, D., (1967) The effects of abstraction and discharges on river water quality. In : River Management (Ed. Issac, P.C.G.) Maclaren and Sons Ltd., London, pp. 153.

Mitra, A.K., (1982) Chemical characteristics of surface waters at a selected gauging station in the river Godavari, Krishna and Tungabhadra. Ind. J. Environ. Hlth. pp. 24(2), pp. 165-79.

Moon, H.P., (1939) Aspects of the ecology of aquatic insects. Trans. Brit. Ent. Soc. 6, pp. 39.

Paterson, C.G. and Nursall, J.R., (1975) The effects of domestic and industrial effluents on a large turbulent river. Water Res. 9, pp. 425-35.

Raina, V., Shah, A.R. and Ahmed, S.R., (1984) Pollution studies on river Jhelum-I : An assessment of water quality. Ind. J. Envrion. Hlth. pp. 26(3), pp. 187-201.

Ram Rao, S.V., Singh, V.P. and Mall, L.P., (1978) Pollution studies of river Khan (Indore) India-I. Biological assessment of pollution. Water Res. 12, pp. 555-59.

Sangu, R.P.S. and Sharma, K.D., (1985) Studies on water pollution of Yamuna river at Agra. Ind. J. Environ. Hlth. 27(3), pp. 257-61.

Sawyer, C.N. and McCarty, P.L. (1967) Chemistry for sanitary engineers. McGraw Hill Book Co., New York, 518 pp.

Saxena, K.L., Chakrabarty, R.N., Khan, A.Q., Chattopadhya, S.N. and Harish Chandra, (1966) Pollution studies of the river Ganges near Kanpur. Environ. Hlth. 8, pp. 270-85.

Sengupta, B., Laskar, S., Das, A.K. and Das, J., (1988) Inorganic pollutants of Ganga water in the region of Berhampore to Katwa, West Bengal. Ind. J. Environ. Hlth. 30(3), pp. 202-08.

Sharma, S., (1988) Potamological studies of the Morar (Kalpi) river, Gwalior, M. Phil Dissertation, Jiwaji University, Gwalior.

Sharma, S., (1991) Pollution studies of Morar (Kalpi) river, Gwalior, India. In : Environmental Series Vol. V. Environmental Contamination and Hygiene. (Arif Ali & Ram Prakash Eds.). pp. 177-83.

Sharma S. Kaushik, S. and Mathur, R., (1989) Pollution studies of Saank, Asaun and Kurai rivers in Madhya Pradesh. In : Management of Aquatic Ecosystems (Eds. V.P. Agrawal, B.N. Desai and S.A.H. Abidi) pp. 211-18.

Shelford, V.E., (1914) An experimental study of the behaviour agreement among animal of an animal community. Biol. Bull. 26, pp. 294.

Upadhyaya, N.P. and Roy, M.N., (1982) Studies in river pollution in Kathmandu valley. Ind. J. Environ. Hlth. 24(2), pp. 124-35.

Vass, K.K., Raina, H.S., Zutshi, D.P. and Khan, M.A., (1977) Hydrobiological studies on river Jhelum. 4(6), pp. 238-42.

Verma, S.R. and Dalela, R.C., (1975) Studies on the pollution of the Kalinadi by industrial waste near Mansurpur. Pt. I : Hydrometric and physicochemical characteristics of the wastes and river water. Acta Hydrochem. Hydrobiol. 8(3), pp. 239-57.

Verma, S.R., (1978) Pollution studies of few rivers of western Uttar Pradesh with reference to biological indices. Proc. Indian Acad. Sci. V. 87B, pp. 123-131.

Verma, S.R., Sharma, P., Tyagi, A., Rani, S., Gupta, A.K. and Dailela, R.C., (1984) Pollution and saprobic status of Eastern Kalinadi. Limnologica, 15 (1), pp. 69-133.

WHO, (1978) Nitrites, Nitrates and N-nitroso Compound. World Health Organization, Environmental Health Criteria-5, W.H.O., Geneva.

36

Remote Sensing and the Pollution Monitoring in the Cauvery River, India

S.M. RAMASAMY,
V. VENKATASUBRAMANIAN
K. JOAN SAM
AND
G. EDWIN CHANDRASEKHAR

Introduction

Due to mammoth growth in population the available water resources on the surface of the earth have become inadequate to cope up with man's needs. These resources too have been facing major onslaughts due to pollution by rapidly growing civilization, industrialization and urbanization, etc.

The art of remote sensing has opened up new vistas in mapping and monitoring the water quality. The study so far conducted reveals that, amongst Landsat Multispectral data, the bands that are having wavelength of 0.8 to 1.1 μm is the best for mapping the water bodies as in the near infrared region of electromagnetic spectrum, the light is totally absorbed and hence the water body appears perfectly black. Studies by Hardy (1981) have proved that the wavelength falling in between 0.4 and 0.7 μm provides good information on the turbidity of the waterbody, as the suspended sediments that are floating on the water bodies cause visible light rediation. Kritokos *et al.*, (1974) have demonstrated that the ratio of visible and near infrared bands can be used for general water quality monitoring. Subsequently, studies have proved that, the thermal infrared photographs contribute better discrimination of the water bodies and the status of the pollutants, as there is a significant difference between the temperature of the pollutant and the adjacent water bodies (Balakrishnan, 1986). From the above studies, it is clear that the water quality monitoring can be done, using remote sensing. Hence, an attempt was made to monitor the water quality in parts of main Cauvery drainage system around Tiruchirapalli region with a prime objective of developing a mathematical model for monitoring the water quality using the existing Thematic mapper bands 1, 2, 3 and 4, IRS 1, 2, 3 and 4 and the SPOT bands, 1, 2 and 3.

Regional Setting

Physiography

The Tiruchirapalli region is located in the centre of Tamil Nadu State. In the north, it is bordered by the hills of Pachaimalai and Kollimalai in the west and Southwest bordered by the uplands and hills belonging to the eastern ghats region and the southeast and east it is bordered by the uplands of mainly Tertiary formations. The drainage system is controlled by the Cauvery river, which branches off into Colleroon in the north and Cauvery in the south, just east of Tiruchirpalli (Fig. 36.1). Number of tributary streams join the Cauvery river both in the northern and southern banks.

Status of Pollution

Tiruchirapalli is one of the major urban centres in Tamil Nadu with the population of 4.28 lakhs and covers an area of 30 sq. km. Number of industries have swung up in and around Tiruchirapalli urban and sub-urban areas. The main industries which are contributing pollution to the Cauvery river are the Ponmali Railway workshop from which the solid wastes are dumped into the Cauvery drainage network, the Tiruchy distilleries which discharges Mollasces in Uyyakondan Canal which is a tributary of the river Cauvery and mainly the Tiruchirapalli domestic sewages which contributes wastes and possible infectious agents to the Cauvery river. In addition, the agricultural fields located on either banks of the Cauvery are also supplying synthetic organic chemicals like detergents and pesticides.

Methodology

Spectro Radiometric Survey

With an objective of developing a model to monitor the surface water quality through remote sensing, the spectro-radiometric survey was conducted for six stations for five days in Cauvery drainage network in and around Tiruchirapalli. The six sites selected are Kamparasampettai, Chinthamani, Sarkarpalayam, near G.H., Pakalarai and Senthaneerpuram. The Kamparasampettai was selected because that was the point where the Cauvery enters into the Tiruchirapalli environs and the Sarkarpalayam was selected because that was the point from which the Cauvery is totally getting out from the Tiruchirapalli environment. Hence, it is presumed that the difference in the pollutants level in between these two stations may give the actual contribution of pollutants from the Tiruchirapalli environment. The Chinthamani and Palakarai were selected as they fall in

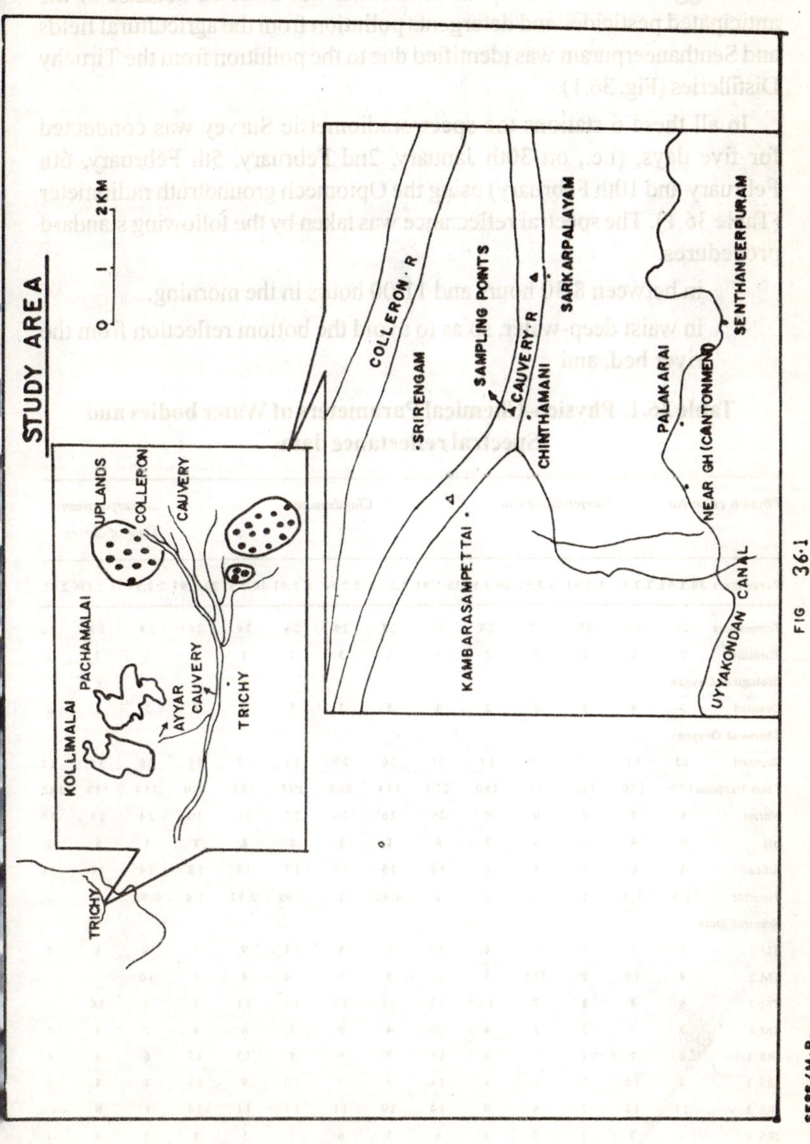

FIG 36·1

SESS/M·R.

the heart of the Tiruchirappali township and bear more sewage from the urban agglomeration. The point near G.H. was selected because of the anticipated pesticides and detergents pollution from the agricultural fields and Senthaneerpuram was identified due to the pollution from the Tiruchy Distilleries (Fig. 36.1).

In all these 6 stations the spectroradiometric Survey was conducted for five days, (i.e., on 30th January, 2nd February, 5th February, 6th February and 10th February) using the Optomech groundtruth radiometer (Table 36.1). The spectral reflectance was taken by the following standard procedures.

- in between 8.30 hours and 11.00 hours in the morning.
- in waist deep-water, so as to avoid the bottom reflection from the river bed, and

Table 36.1: Physico-Chemical Parameters of Water bodies and Spectral reflectance data

Physico chemical	*Kamparasampetai*					*Chinthamani*					*Sarkarpalayam Date of Survey*			
Parameters	**30.1.91**	**2.2.91**	**5.2.91**	**6.2.91**	**10.2.91**	**30.1.91**	**2.2.91**	**5.2.91**	**6.2.91**	**10.2.91**	**30.1.91**	**2.2.91**	**5.2.91**	**6.2.91**
Temperature	28	29	29	27	28	28	28	28	26	29	26	28	27	26
Turbidity	2	2	2	2	2	3	3	3	3	3	3	3	3	3
Biological Oxygen Demand	3	4	2	2	2	8	5	7	5	4	4	5	5	6
Chemical Oxygen Demand	13	11	11	10	11	31	26	29	28	27	35	26	31	33
Total Hardness	190	170	162	56	160	220	184	208	191	188	190	184	188	188
Nitrate	1	2	0	0	0	29	26	29	27	26	30	24	25	30
pH	9	9	9	9	9	8	8	8	8	8	8	8	8	8
Silica	6	6	6	5	6	18	16	17	17	18	18	14	16	18
Fluoride	1.3	1.3	1	1	1	1	0.96	1	0.99	0.97	1.4	0.92	1	1
Spectral Data														
TM 1	8	7	5	5	6	13	8	8	13	9	7	6	8	8
TM 2	8	10	8	715	9	10	9	9	10	8	9	10		
TM 3	9	8	8	7	11	17	11	11	11	11	8	9	10	
TM 4	3	3	2	1	4	7	4	6	5	6	4	2	4	4
IRS 1	6	7	4	6	8	11	7	9	8	13	13	6	8	6
IRS 2	8	10	7	6	8	14	9	9	10	9	18	7	8	9
IRS 3	11	10	7	6	9	14	10	11	11	11	24	8	9	10
IRS 4	4	3	2	3	3	6	5	6	5	5	8	3	4	4
SPOT 1	9	7	6	6	6	13	8	6	6	8	43	7	9	9
SPOT 2	12	8	7	6	7	14	9	11	10	9	32	8	11	10
SPOT 3	3	3	2	1	2	6	4	5	5	4	9	2	4	4

Table-1 (Contd.)

				Near GH					*Palakrai*					*Senthaneepuram*	
10.2.91	10.1.91	2.2.91	5.2.91	6.2.91	10.2.91	30.1.91	2.2.91	5.2.91	6.2.91	10.2.91	30.1.91	2.2.91	5.2.91	6.2.91	10.2.91
29	29	29	29	29	30	29	30	30	29	28	28	28	28	28	28
3	8	7	7	7	6	8	8	8	7	8	3	4	4	4	4
5	31	30	25	24	21	29	31	26	20	34	5	6	6	5	6
28	94	91	87	85	75	92	97	99	89	81	98	22	24	25	24
184	228	206	196	200	164	217	229	210	201	240	190	200	198	204	206
26	16	15	12	15	10	156	17	15	14	18	10	15	12	10	15
8	8	8	8	8	8	8	8	8	8	8	8	8	8	8	8
16	21	18	14	16	19	21	21	29	18	22	8	19	19	19	19
0.94	1.18	1.16	1.15	1.14	1.09	1.15	1.16	1.13	1.19	0.84	1.17	1	1	0.84	
7	5	7	8	8	7	8	7	7	8	11		7	8	6	9.8
5	6	8	7	8	8	9	9	6	9	13	9	6	7	10	7
5	7	8	8	10	8	9	11	6	8	13	11	6	8	11	10
3	4	4	3	4	5	5	5	4	4	8	2	3	3	4	7
7	5	6	7	8	8	8	7	5	9	15	6	7	7	7	10
9	5	7	9	9	8	8	8	5	8	14	7	6	11	8	8
9	4	10	9	9	8	7	8	8	8	13	7	6	8	10	9
4	2	4	5	5	5	4	6	5	4	4	8	4	4	6	3.7
9	10	7	8	9	8	9	8	6	7	14	4	6	11	8	8
9	1	8	9	10	8	17	9	5	7	16	7	6	8	9	10
3	1	3	4	4	5	4	6	3	3	8	4	4	4	3	5
3	1	3	4	4	5	4	6	3	3	8	4	4	4	3	5

– by using the normal procedures adopted in the ground radiometric survey.

Water Quality Analysis

On the spot measurement

Immediately after the collection of the radiometric data from the particular spot, the temperature and pH were measured on the spot and the water samples were collected in two bottles, one with Nitrogen fixation and the other without the same for carrying out the chemical analysis in the laboratory.

Laboratory analysis

In the laboratory the parameters—Biological Oxygen demand, chemical oxygen demand, total hardness, pH, nitrate, silica and fluoride were estimated and the 11 band spectral reflectance data one to one correlation was made in between the Physico chemical parameters and the spectral data by both the graphical method and also by bivariate regression analysis using the mini vax computer system.

Results and Discussion

Graphical method

Temperature Versus Spectral data : The 30 temperature values observed for 5 days from six stations show that the temperature is ranging from 26 to 30°C. The correlation of temperature data with the respective 11 and spectral data (Fig. 36.2) shows that the temperature has got a positive correlation with the spectral bands of TM2, TM4, IRS2, IRS3 and SPOT 1 (Table 36.2). It shows that the spectral reflectance value increases with an increase of temperature of the water body in the above bands. Amongst these bands the TM2 and IRS2 bands seem to be best suited for temperature monitoring as positive correlation is observed in two stations. (Table 36.3)

Table 36.2 : Correlation between Physico-Chemical Parameters and Special bands

Physico - Chemical Parameters	*KAMPARASAMPETAI*										
	TM bands				*IRS bands*				*Spot bands*		
	1	2	3	4	1	2	3	4	1	2	3
Temperature	F	⊕	F	F	F	⊕	F	N	F	F	F
Turbidity	Θ	⊕	Θ	N	F	⊕	N	N	N	N	N
Biological Oxygen Demand	N	⊕	F	F	N	⊕	N	N	N	F	N
Chemical Oxygen Demand	–	–	–	–	–	–	–	–	–	–	–
Total Hardness	F	F	F	⊕	F	F	F	F	⊕	⊕	⊕
Nitrate	–	–	–	–	–	–	–	–	–	–	–
pH	–	–	–	–	–	–	–	–	–	–	–
Silica	N	⊕	N	N	N	⊕	N	N	N	N	N
Flouride	–	–	–	–	–	–	–	–	–	–	–

⊕ Positive Correlation

F Fair Correlation

– Negative Correlation

N No Correlation

Table 36.2 (Contd.)

Table 36.2 (Contd.)

Physico - Chemical Parametes	*TM bands*				*Chinthamani IRS bands*				*Spot bands*		
	1	2	3	4	1	2	3	4	1	2	3
Temperature	F	N	N	N	N	N	N	N	N	N	N
Turbidity	–	–	–	–	–	–	–	–	–	–	–
Biological Oxygen Demand	F	⊕	F	F	F	⊕	N	⊕	N	⊕	⊕
Chemical Oxygen Demand	F	⊕	F	F	F	⊕	N	⊕	N	⊕	⊕
Total Hardness	F	⊕	F	F	F	⊕	N	⊕	N	⊕	⊕
Nitrate	N	⊕	N	F	F	⊕	N	⊕	N	⊕	N
pH	–	–	–	–	–	–	–	–	–	–	–
Silica	N	N	N	N	N	⊕	F	N	N	N	N
Flouride	–	–	–	–	–	–	–	–	–	–	–

Table 36.2 (Contd.)

Turbidity Versus Spectral data : The turbidity varies from 2-8.5 nepheloturbidity units and it is minimum in Kamparasampattai, where the pollution seems to be less and more in Palakarai where the domestic sewages are dumped to the maximum. The profiling of the turbidity values and the respective 11 band spectral data (Fig. 36.3) shows that the turbidity has a positive correlation mostly with the bands of TM2 and IRS2 (Table 36.2). It indicates that the turbidity variations can be monitored in these spectral ranges.

Biological Oxygen demand Versus Spectral data : The Biological oxygen demand ranges from 2 to 34 mg/l and once again the minimum at Kamparasampattai and the maximum in Palakarai. The profiling of BOD data for different stations and the respective profiles of the 11 band spectral data (Fig. 36.4) shows that the band TM2 seems to be the most suitable band for monitoring BOD. That is when BOD is more in water, the spectral reflectance is more in TM band 2. Further the analysis shows that IRS1, IRS2, IRS4, SPOT1 and SPOT2 bands also seem to have better

Table 36.2 (Contd.)

Physico - Chemical Parameters	*Sarkarpalayam*										
	TM bands				*IRS bands*				*Spot bands*		
	1	2	3	4	1	2	3	4	1	2	3
Temperature	F	F	F	F	F	⊕	⊕	F	⊕	F	F
Turbidity	–	–	–	–	–	–	–	–	–	–	–
Biological Oxygen Demand	F	⊕	F	F	F	F	F	F	F	F	F
Chemical Oxygen Demand	F	F	⊕	F	F	F	F	F	F	F	F
Total Hardness	F	F	⊕	F	F	F	F	F	F	F	F
Nitrate	F	⊕	⊕	F	F	⊕	F	⊕	F	F	F
pH	–	–	–	–	–	–	–	–	–	–	–
Silica	F	⊕	⊕	F	F	⊕	F	⊕	F	F	F
Flouride	F	⊕	⊕	F	F	⊕	F	⊕	F	F	F

Table 36.2 (Contd.)

positive correlation with BOD, but the TM2 band (Table 36.3) is the best suited band for monitoring the BOD as TM 2 shows such a relation in 3 stations (Tables 36.2 and 36.3).

Chemical Oxygen Demand Versus Spectral data : The COD values varies from 10.24 to 98.16 mg/l. The lowest value is recorded from Kamparasampettai and the highest from Palakarai. The graphical correlation of the COD data and the 11 band spectral data (Fig. 36.5) shows that the COD is having a positive correlation with TM2, TM3, IRS2, IRS4, SPOT2 and SPOT3 band (Table 36.2) suggesting an increase in reflectance with increase in COD values.

However, amongst all the bands, the IRS4 band seems to be the best suited band for COD monitoring as it shows positive correlation in 4 stations and followed by TM2 and SPOT2 (Table 36.3), whereas the TM3, TM4, IRS2, SPOT1 and SPOT3 bands shows positive correlation only in one station each (Tables 36.2 and 36.3).

Total hardness Versus Spectral data : The total hardness value analysed for the water samples varies from 156-240 mg/l, the minimum in Kamparasampettai and the maximum in Palakarai. The correlation of total

Table 36.2 (Contd.)

Physico - Chemical Parameters	TM bands				Sarkarpalayam IRS bands				Spot bands		
	1	2	3	4	1	2	3	4	1	2	3
Temperature	N	@	F	@	N	N	N	N	N	N	N
Turbidity	–	–	–	–	–	–	–	–	–	–	–
Biological Oxygen Demand	N	N	N	N	C	N	@	@	F	F	F
Chemical Oxygen Demand	F	F	F	F	F	F	F	@	F	F	F
Total Hardness	F	F	N	N	N	N	N	F	@	C	N
Nitrate	F	N	N	N	N	N	N	N	N	N	C
pH	–	–	–	–	–	–	–	–	–	–	–
Silica	F	N	N	N	N	N	N	N	N	N	C
Flouride	–	–	–	–	–	–	–	–	–	–	–

Table 36.2 (Contd.)

hardness with 11 band spectral data (Fig. 36.6) shows that it has a positive correlation mostly with the bands of SPOT and IRS spectral channels (Table 36.2). SPOT 1 seems to be the most suitable band for monitoring the total hardness, as it is repeated in 4 stations in the analysis (Table 36.3). The other suitable bands are TM2, TM3, IRS4, SPOT2 and SPOT3 (Table 36.3).

Nitrate Versus Spectral data : The analysis of 30 water samples shows that the Nitrate is varying from 1.5 to 30 mg/l, the former from Kamparasampettai and the later from Sarkarpalayam. Its comparison with the spectral data (Fig.36.7) shows that the TM band 2 seems to be the best suited band for monitoring the nitrate, whereas the other bands like TM2, TM3 and IRS4 come next in the hierarchy (Tables 36.2 and 36.3).

pH Versus Spectral data : The pH value doesn't show much change in the most of the stations except in Senthaneerpuram. The analysis of Senthaneerpuram samples (Fig. 36.8) shows that TM2, TM3, IRS1, and IRS3 seems to have a fair correlation with pH data (Tables 36.2 and 36.3).

Silica Versus Spectral data : The silica content varies from 5.5 to 22 mg/l, less in Kamparasampettai and more in Palakarai. The graphical

Table 36.2 (Contd.)

Physico - Chemical Parameters					*Palakarai*						
	TM bands				*IRS bands*				*Spot bands*		
	1	2	3	4	1	2	3	4	1	2	3
Temperature	N	F	F	F	F	F	F	F	F	F	F
Turbidity	F	⊕	⊕	N	N	N	N	N	N	N	N
Biological Oxygen Demand	F	⊕	⊕	N	F	F	F	⊕	⊕	F	F
Chemical Oxygen Demand	F	⊕	⊕	N	F	F	F	⊕	⊕	F	F
Total Hardness	F	⊕	⊕	N	F	F	F	⊕	⊕	F	F
Nitrate	N	N	N	N	N	N	⊕	⊕	N	N	⊕
pH	–	–	–	–	–	–	–	–	–	–	–
Silica	N	N	N	N	N	N	⊕	⊕	N	N	⊕
Flouride	–	–	–	–	–	–	–	–	–	–	–

Table 36.2 (Contd.)

analysis shows that TM2, TM3, IRS2, IRS3, IRS4 and SPOT3 have a good correlation with the Silica values (Fig. 36.9). Amongst all the bands the IRS2 is the most suitable band, as it shows positive correlation in 3 stations (Table 36.3).

Fluoride Versus Spectral data : The fluoride content varies from 0.84 to 1.4 mg/l and it is the minimum in Senthaneepuram and maximum in Sarkarpalayam samples. As a whole there is no much variation in fluoride content (Fig. 36.10). However, the analysis of Sarkarpalayam samples shows appreciable variations and the spectral bands TM2, IRS2 and IRS4 seem to have a positive correlation (Fig. 36.10, Tables 36.2 & 36.3).

Regression Analysis

Subsequent to the graphical methods, the bivariate regression analysis was attempted by stacking all the Physico-Chemical parameters data and the spectral data. The Physico-Chemical parameters were kept as dependent variables and the spectral data as independent varaibles. For example, in the first stage the temperature data was taken as the dependent variable and it was correlated with the respective 11 band spectral data.

Table 36.2 (Contd.)

Physico - Chemical Parameters	*TM bands*				*Senthannerpuram IRS bands*				*Spot bands*		
	1	2	3	4	1	2	3	4	1	2	3
Temperature	N	N	N	N	N	N	N	N	N	N	N
Turbidity	@	N	C	F	@	N	N	N	N	N	F
Biological Oxygen Demand	F	F	C	F	@	N	C	N	N	N	N
Chemical Oxygen Demand	–	–	–	–	–	–	–	–	–	–	–
Total Hardness	N	N	N	N	@	N	N	N	N	N	N
Nitrate	N	C	C	N	@	N	C	N	N	N	N
pH	N	C	C	N	@	N	@	N	C	N	N
Silica	N	C	C	N	@	N	C	N	N	N	N
Flouride	–	–	–	–	–	–	–	–	–	–	–

As per the ethics of the bivariate regression analysis, the temperature data of 30 samples was correlated with respective TM1 data and the best fitting graph was drawn in such a way that $y = mx + c$. From such linear correlation matrix/graph, coefficient A, Coefficient B, Correlation coefficient, standard error, T Value, and F value were worked out. Similarly these values were worked from all the other 10 graphs deduced from

$y = mx + c$ graph drawn for temperature versus other 10 bands

The worked out correlation coefficient data between the temperature and other spectral data were tabulated in descending order and the top 3 bands were selected as the best bands showing good correlation with the temperature (Table 36.4). In the same way the best spectral bands amongst 11 spectral bands which are showing good correlation with turbidity, BOD, COD, total hardness, nitrate, pH, silica and fluoride were also tabulated (Table 36.4).

Subsequently, the mathematical equations were developed for working out the different physico-chemical parameters using the spectral band which are showing maximum correlation coefficient (Table 36.4). As per the bivariate regression analysis, the temperature of a water body can be worked out by the following formula.

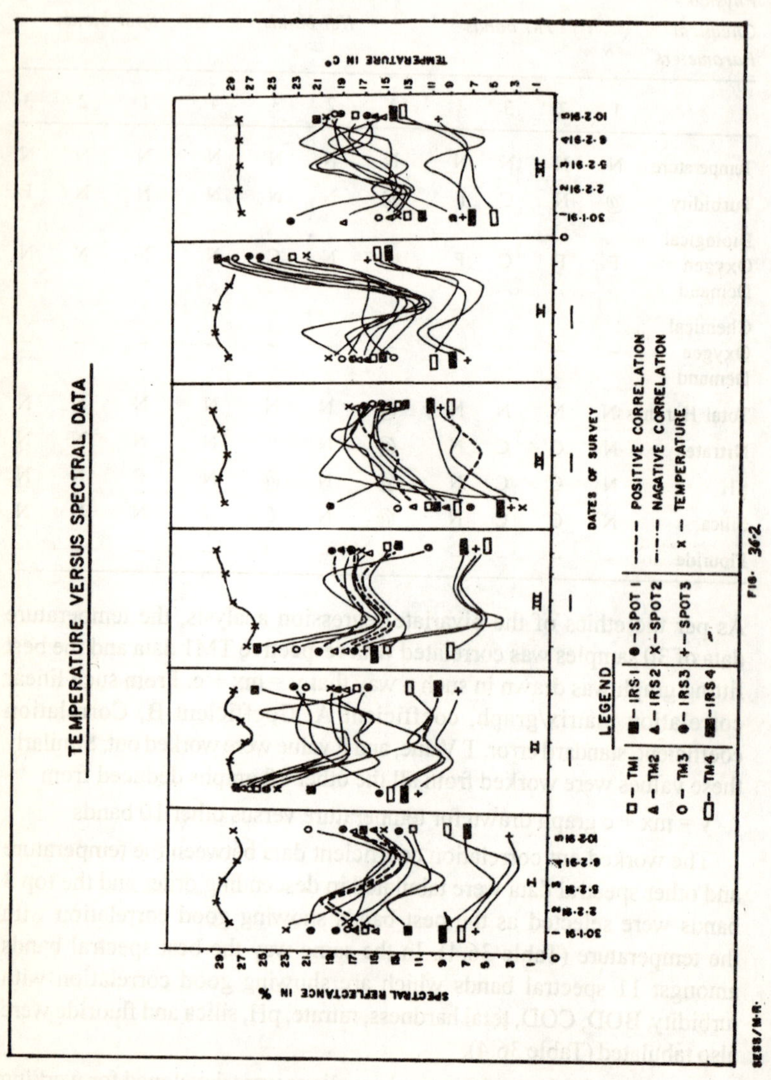
TEMPERATURE VERSUS SPECTRAL DATA
LEGEND
□ — TM1 ■ — IRS 1 ● — SPOT 1 ——— POSITIVE CORRELATION
△ — TM2 ▲ — IRS 2 X — SPOT 2 -·-·- NAGATIVE CORRELATION
O — TM3 ● — IRS 3 + — SPOT 3 X — TEMPERATURE
▯ — TM4 ■ — IRS 4
DATES OF SURVEY
SPECTRAL REFLECTANCE IN %
TEMPERATURE IN C°
FIG- 36.2

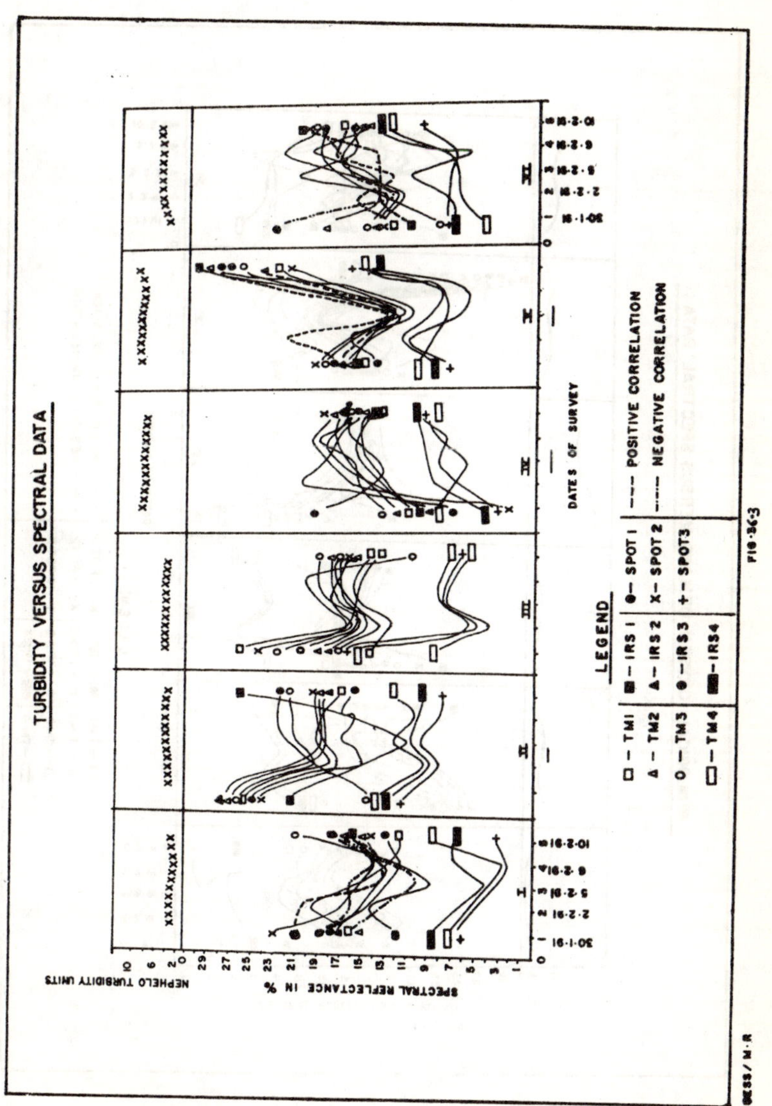
TURBIDITY VERSUS SPECTRAL DATA
NEPHELO TURBIDITY UNITS
SPECTRAL REFLECTANCE IN %
DATES OF SURVEY
30-1-91
2-2-91
5-2-91
6-2-91
10-2-91
LEGEND
□ – TM1
Δ – TM2
O – TM3
▭ – TM4
■ – IRS 1
▲ – IRS 2
● – IRS 3
▬ – IRS4
⊕ – SPOT 1
X – SPOT 2
+ – SPOT3
POSITIVE CORRELATION
NEGATIVE CORRELATION

FIG·36·3

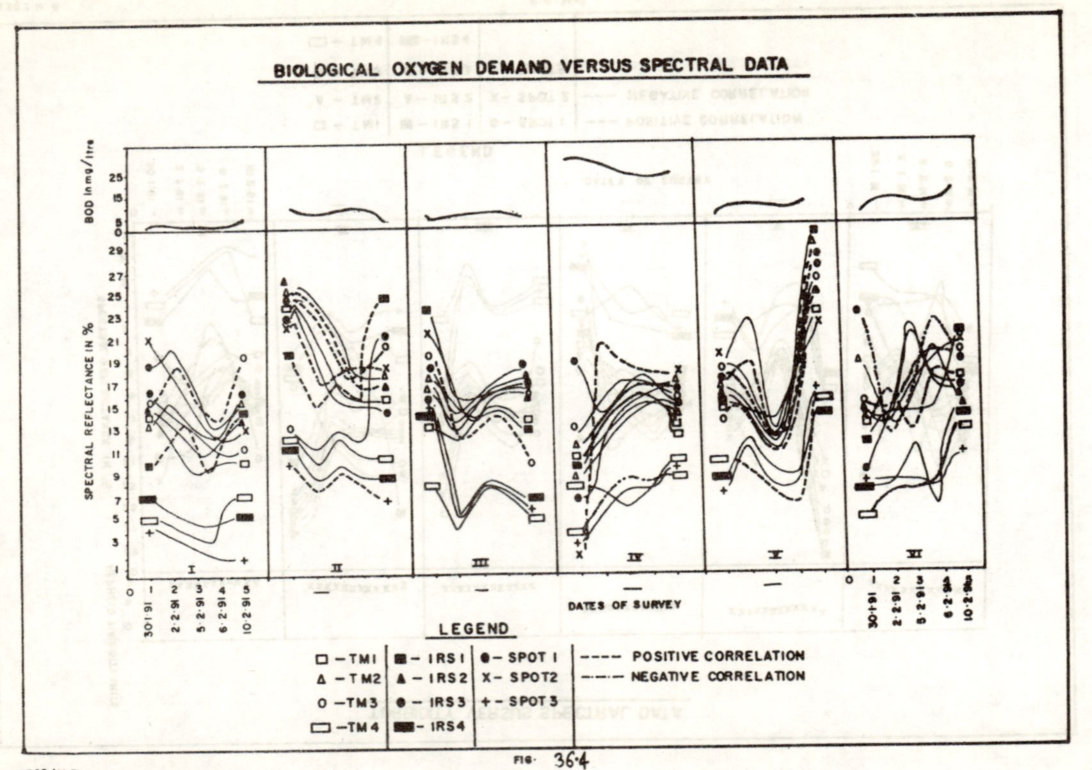
BIOLOGICAL OXYGEN DEMAND VERSUS SPECTRAL DATA
BOD in mg/litre
SPECTRAL REFLECTANCE IN %
DATES OF SURVEY
30.1.91
2.2.91
5.2.91
6.2.91
10.2.91
I
II
III
IV
V
VI
LEGEND
□ – TM1
Δ – TM2
O – TM3
▭ – TM4
■ – IRS1
▲ – IRS2
● – IRS3
▬ – IRS4
● – SPOT 1
X – SPOT2
+ – SPOT3
---- POSITIVE CORRELATION
-·-·- NEGATIVE CORRELATION
SESS/M.R

FIG. 36.4

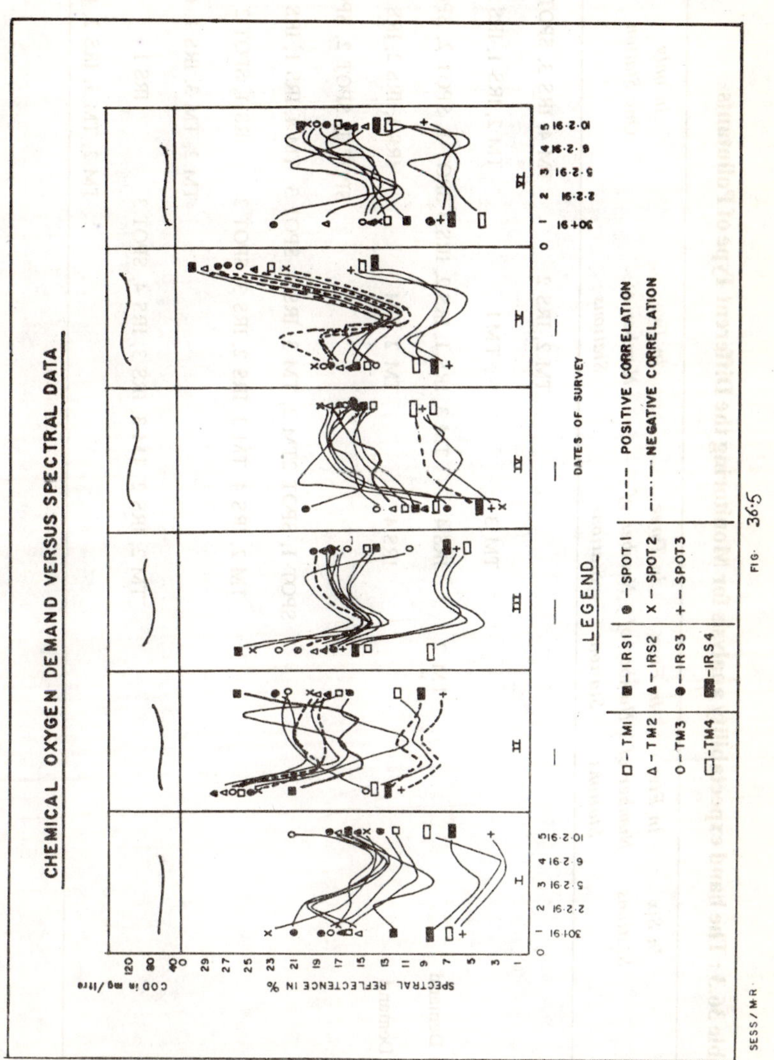
CHEMICAL OXYGEN DEMAND VERSUS SPECTRAL DATA
COD in mg/litre
SPECTRAL REFLECTENCE IN %
DATES OF SURVEY
30·1·91
2·2·91
5·2·91
6·2·91
10·2·91
I
II
III
IV
V
VI
LEGEND
□ – TM1
Δ – TM2
O – TM3
▭ – TM4
■ – IRS1
▲ – IRS2
● – IRS3
▬ – IRS4
⊗ – SPOT 1
X – SPOT 2
+ – SPOT 3
---- POSITIVE CORRELATION
–·–·– NEGATIVE CORRELATION
FIG. 36·5
SESS/MR

Table 36.3 : The band expectability analysis for Monitoring the Different Type of Pollutants

Physico-Chemical Parameters	*In Six Stations*	*In Five Number of Stations*	*In Four Number of Stations*	*In Three Number of Stations*	*In Two Number of Stations*	*In only One Station*
Temperature					TM 2, IRS 2	TM 4, IRS 3, SPOT 1
Turbidity				TM 3	TM1	TM 2, IRS 1, IRS 2
Biological Oxygen Demand			TM 2	IRS 4	TM 3, IRS 1, IRS 2, IRS 3	SPOT 1, SPOT 2, SPOT 3
Chemical Oxygen Demand				IRS 4	TM 2, TM 3	IRS 1, IRS 2, IRS 3
						SPOT 1, SPOT 2, SPOT 3
Total Hardness				SPOT 1, SPOT 2	TM 2, TM 3, IRS 4, SPOT 3	TM 4, IRS 1, IRS 2
Nitrate				TM 2, IRS 4	TM 3, IRS 2, IRS 3, SPOT 2	IRS 1, SPOT 2
pH						TM 2, TM 3, IRS 1, IRS 3
Silica				TM 2, IRS 2	TM 3, IRS 3, IRS 4, SPOT 3	IRS 1
Flouride						TM 2, TM 3, IRS 2, IRS 4

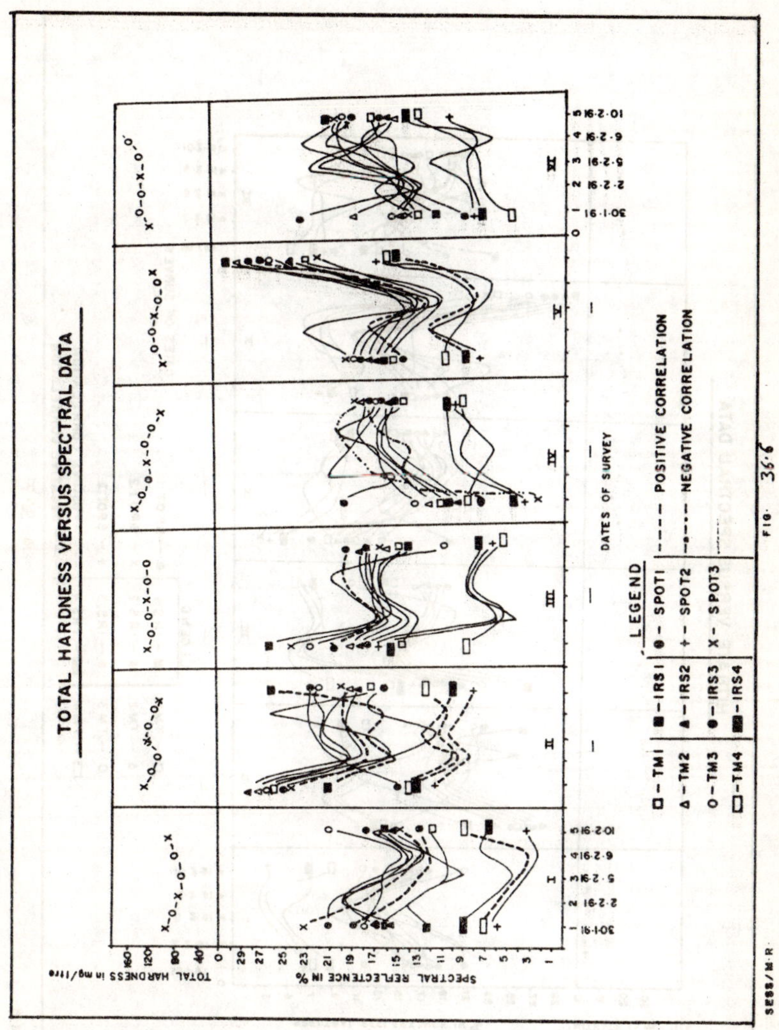
TOTAL HARDNESS VERSUS SPECTRAL DATA
TOTAL HARDNESS in mg/litre
SPECTRAL REFLECTENCE IN %
DATES OF SURVEY
30-1-91
2-2-91
5-2-91
6-2-91
10-2-91
I
II
III
IV
V
VI
LEGEND
□ - TM1
Δ - TM2
O - TM3
▭ - TM4
■ - IRS1
▲ - IRS2
● - IRS3
▬ - IRS4
⊕ - SPOT1
+ - SPOT2
X - SPOT3
POSITIVE CORRELATION
NEGATIVE CORRELATION
SEBB/M.R
FIG. 36.6

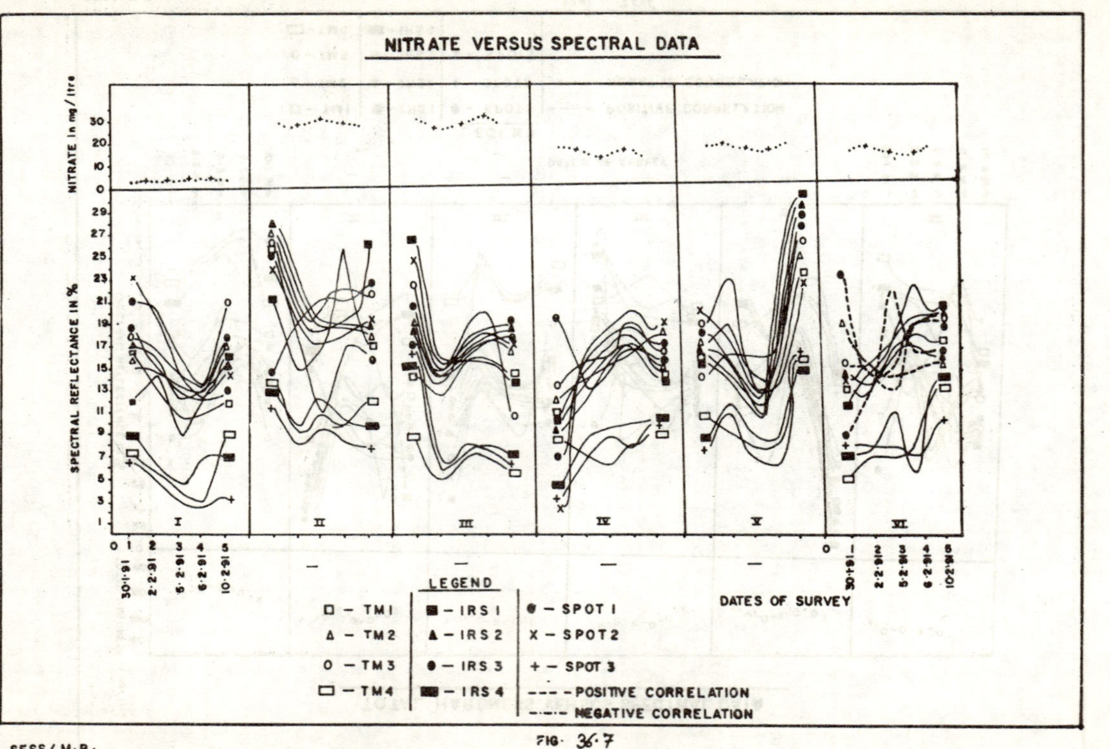
NITRATE VERSUS SPECTRAL DATA
NITRATE in mg/litre
SPECTRAL REFLECTANCE IN %
I
II
III
IV
V
VI
30·1·91
2·2·91
5·2·91
6·2·91
10·2·91
DATES OF SURVEY
LEGEND
□ – TM1
Δ – TM2
O – TM3
□ – TM4
■ – IRS 1
▲ – IRS 2
● – IRS 3
■ – IRS 4
⊕ – SPOT 1
X – SPOT 2
+ – SPOT 3
---- POSITIVE CORRELATION
-·-·- NEGATIVE CORRELATION
SESS/M.R.

FIG. 36.7

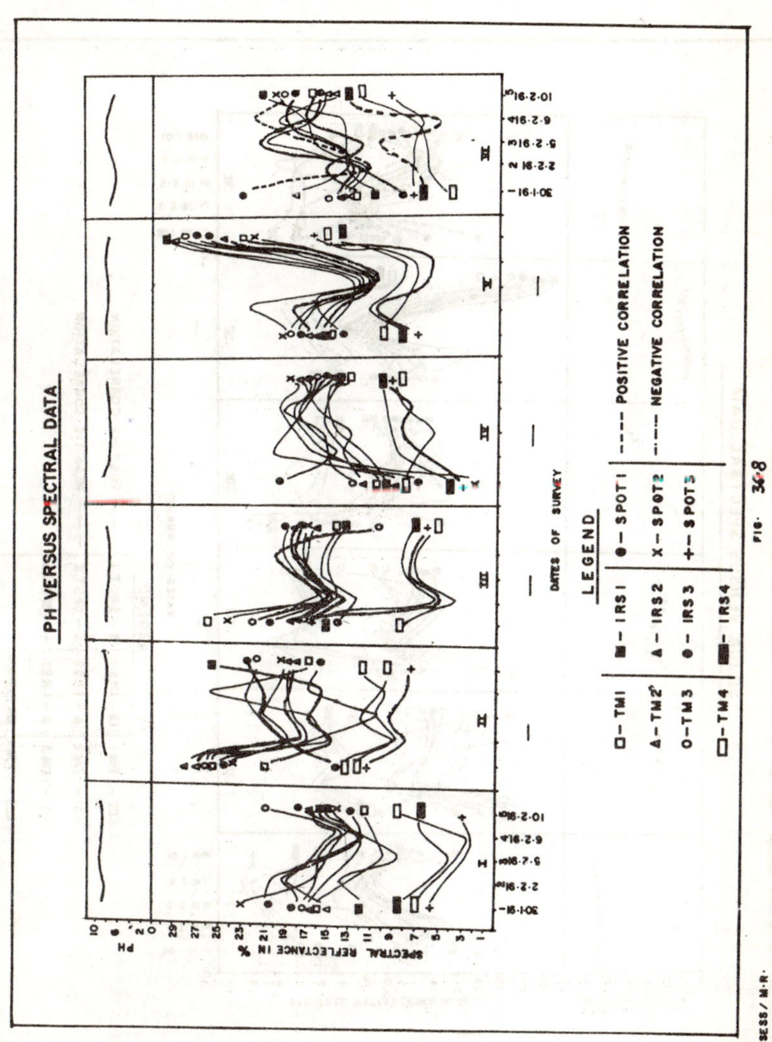
PH VERSUS SPECTRAL DATA
PH
SPECTRAL REFLECTANCE IN %
30·1·91
2·2·91
5·2·91
6·2·91
10·2·91
I
II
III
IV
V
VI
DATES OF SURVEY
LEGEND
□ – TM1
Δ – TM2
O – TM3
□ – TM4
■ – IRS1
Δ – IRS2
● – IRS3
■ – IRS4
● – SPOT1
X – SPOT2
+ – SPOT3
---- POSITIVE CORRELATION
–·–·– NEGATIVE CORRELATION
SESS / M·R·

FIG. 36·8

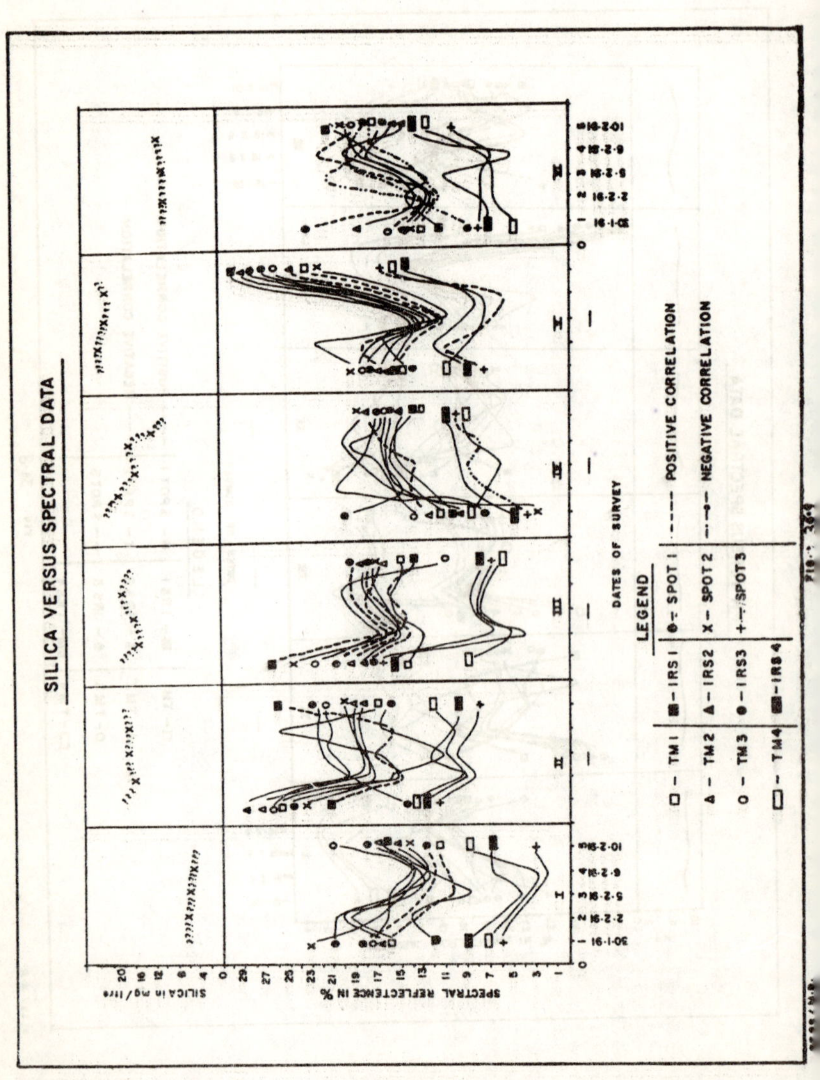
SILICA VERSUS SPECTRAL DATA
SILICA in mg/litre
SPECTRAL REFLECTENCE IN %
DATES OF SURVEY
LEGEND
POSITIVE CORRELATION
NEGATIVE CORRELATION

Table 36.4 : Results of Bivariate Regression Analysis

Dependent variable	*Independent variable*	*Coefficient A*	*Coefficient B*	*Correlation Coefficient*	*Standard Error*	*T-value*	*F-Value*
	SPOT-2	29.26228	-9.6594	0.4399	1.0147	2.5925	6.72
Temperature	IRS-2	29.81237	-0.1308	0.4061	1.0325	2.3515	6.52
	IRS-3	29.57625	-0.1469	0.4073	1.0319	2.3604	5.57
	IRS-4	2.9899	-0.2236	0.2236	2.2578	1.2140	1.47
Turbidity	SPOT-3	3.4309	-0.2919	0.2166	2.2614	1.1744	1.37
BOD	IRS-4	4.7196	1.6148	0.1250	10.9693	1.1652	1.35
COD	IRS-4	21.5195	5.1958	0.2375	31.7870	1.2938	1.67
	SPOT-3	27.8685	4.4480	0.2336	31.8171	1.2716	1.61
	SPOT-3	173.6270	5.4222	0.4465	18.6762	2.6409	6.97
Total Hardness	IRS-4	166.9600	6.1010	0.4372	7723	2.5725	6.61
	TM-1	158.4457	4.6859	0.4319	18.8255	2.8901	6.42
	SPOT-3	4.4702	2.9557	0.5238	8.2631	3.2538	10.5
Nitrate	IRS-1	1.2719	1.8791	0.4719	8.5521	2.8326	8.02
	TM-1	2.7282	2.4166	0.4793	8.5133	2.8901	8.35
pH	SPOT-3	8.3513	6.1803	0.4660	0.2016	2.7874	7.77
	IRS-4	8.5706	6.1597	0.4042	0.2085	2.3383	5.46
	SPOT-3	8.6496	1.4350	0.4537	4.8454	2.6940	7.25
Silica	IRS-4	7.1453	1.5582	0.4286	4.9123	2.5108	6.30
	SPOT-1	0.9698	1.0165	0.5403	0.1078	3.3972	11.5
Flouride	SPOT-2	0.9649	1.0141	0.4073	0.1171	2.3600	5.56

provided the spectral data is known for that water body of that particular area.

Temperature = Coefficient A ± (Coefficient B × spectral data)

Hence the temperature of a water body can be worked out as follows :

1. Temperature = 29.26228 – (.96594 × SPOT 2) if SPOT 2 data is available.
2. Temperature = 29.81237 – (0.1308 × IRS 2) if IRS 2 data is available.
3. Temperature = 29.57657 – (0.1469 × IRS 3) if IRS 3 data is available.

Table 36.5 : Mathematical Models for Monitoring Different Pollutants

Parameter	*Coefficient-A*	*± Coefficient-B*	*× Spectral Data*
Temperature	29.26228	-9.6594 E-02	SPOT 2
	29.81237	-0.1308	IRS 1
Turbidity	2.9899	0.3460	IRS 4
BOD	4.7196	1.61488	IRS 4
COD	21.5195	5.1958	IRS 4
Total Hardness	173.6270	5.4222	SPOT 3
	166.9604	6.1010	IRS 4
Nitrate	4.4702	2.9527	SPOT 3
	1.2719	1.8791	IRS 1
pH	8.5313	-6.1830 E-02	SPOT 3
Silica	8.6496	1.4350	SPOT 3
Fluoride	0.9698	1.163 E-02	SPOT 1

Thus, the temperature of a given water body can be worked out provided SPOT 2, IRS 2 and IRS 3 data are available. But this model has got certain limitations, due to the variation in the spectral signature of a particular terrestrial object from place to place depending upon the terrain conditions, sun azimuth and inclination of the sun. However, this model will be valid at least for the regions which are falling with the same elevation and terrain conditions with which the azimuth and sun elevation angle are same.

One interesting feature observed in the graphical method for the temperature value is TM2, TM4, IRS2, IRS3 and SPOT1 bands show positive correlation, whereas in the bivariate regression analysis it shows maximum correlation with the bands SPOT2, IRS2 and IRS3. This is because

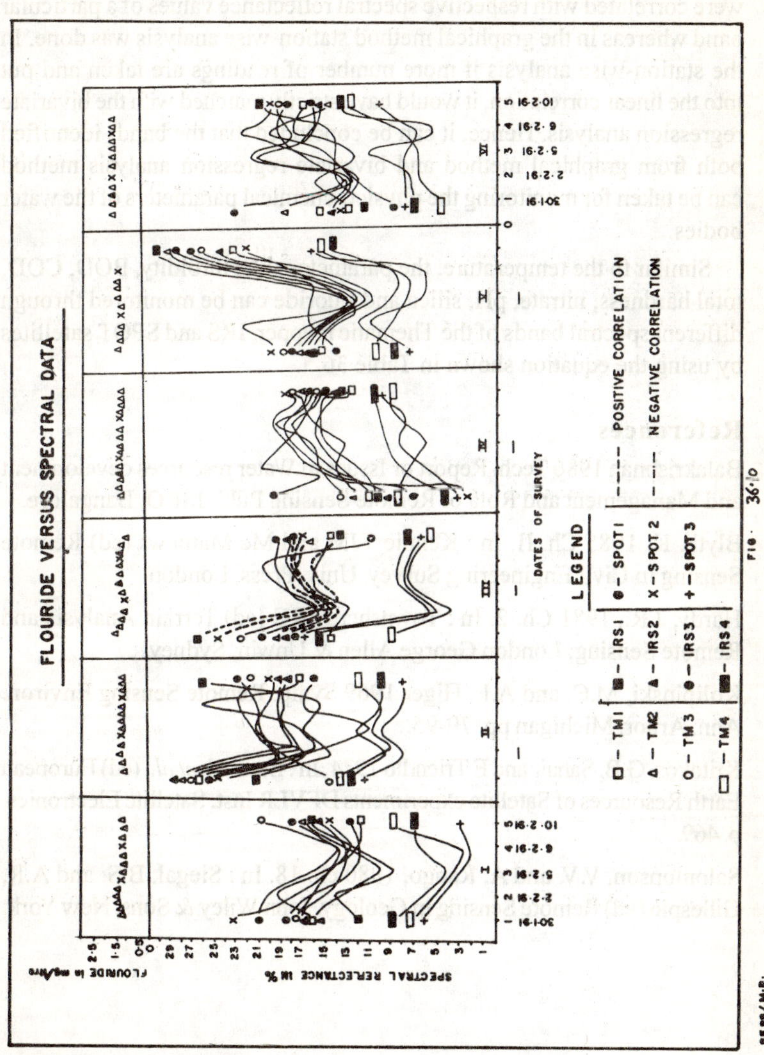
FLOURIDE VERSUS SPECTRAL DATA
FLOURIDE in mg/litre
SPECTRAL REFLECTANCE IN %
DATES OF SURVEY
LEGEND
□ – TM1
△ – TM2
O – TM3
▯ – TM4
■ – IRS1
▲ – IRS2
● – IRS3
▬ – IRS4
● – SPOT 1
X – SPOT 2
+ – SPOT 3
—— POSITIVE CORRELATION
– – – NEGATIVE CORRELATION
SESS/M.R.

FIG. 36·10

of the fact, that in bivariate regression analysis all the temperature values were correlated with respective spectral reflectance values of a particular band whereas in the graphical method station-wise analysis was done. In the station-wise analysis if more number of readings are taken and put into the linear correlation, it would have strictly matched with the bivariate regression analysis. Hence, it can be concluded that the bands identified both from graphical method and bivariate regression analysis method can be taken for monitoring the physico-chemical parameters of the water bodies.

Similar to the temperature, the parameters like turbidity, BOD, COD, total hardness, nitrate, pH, silica and fluoride can be monitored through different spectral bands of the Thematic mapper, IRS and SPOT satellites by using the equation shown in Table 36.5.

References

Balakrishnan 1986 Tech. Report or Issues in Water resources development and Management and Rola of Remote Sensing Publ. ISRO, Bangalore.

Blyth. K. 1985 Ch. II. In : Kennie TJM and Mc Mathews (ed) Remote Sensing in Civil Engineering Survey. Univ. Press, London.

Hardy, J.R., 1981 Ch. 2. In : Townsherd JRG (ed) Terrain Analysis and Remote Sensing, London George Allen & Unwin, Sydney.

Kolipinski, M.C. and A.L. Higer 1969 Symp. Remote Sensing Environ. Ann. Arbor, Michigan pp. 79-95.

Kritokos, G.B. Sahair and E Triendlo 1974. In : Battrock *et al.*, (ed) European Earth Resources of Satellite experiments DFVLR Inst. Satellite Electronics, p. 469.

Salomonson, V.V. and A. Rango, 1980 ch. 18. In : Siegal, B.S. and A.R. Gillespie (ed) Remote Sensing in Geology. John Wiley & Sons, New York.

37

Toxicity of Fenvalerate to the Fresh Water Fishes *Labeo rohita* (Hamilton) and *Aplocheilus panchax* (Hamilton)

K.S. TILAK,
A. ANNAMANI,
*S. JHANSI LAKSHMI
AND
T. ANITA SUSAN

Introduction

In agriculture, the application of pesticides is inevitable. The use of highly potent and photoable synthetic pyrethroid pesticides have been on the increase during the last decade (Elliott, 1977). Among the pyrethroids, the neuro-toxic organic insecticide fenvalerate (-cyano (3-Phenoxy Phenyl)-methyl 4 chloro-(I. methyethyl benzene acetate) is used extensively for the control of boll weevil, *Anthonomous grandis* (Boheman) and boll worm, *Heliothis zea* (Boddie) on cotton, tobacco plants and vegetables (Southwick *et al.*, 1983; Talekar *et al.*, 1983). The transportation of the pesticides into aquatic environment via different agencies poses the problem of contamination and mostly the non-target organisms are often affected. Studies were conducted on fishes using the synthetic pyrethroid, fenvalerate (Zitko *et al.*, 1977, Mulla *et al.*, 1978, Coats & O-Donnel Jefferey, 1979, Zitko *et al.*, 1979, Anderson, 1982, Bradbury *et al.*, 1984 and Chapman *et al.*, 1985).

The present investigation was undertaken to determine the toxicity of fenvalerate using static and continuous flow-through systems for two different freshwater fishes, *Labeo rohita* and *Aplocheilus panchax*. The residue analysis for qualitative confirmation was also done to assess the deposition of pesticide in different tissues.

Materials and Methods

Labeo rohita, 2.5 to 4.0 cm and 300-400 mg were brought from a local fish farm and *Aplocheilus panchax* of same size and weight were collected from a nearby irrigation canal at the University campus. The fish were acclimatized to the laboratory conditions in well aerated unchlorinated tap water at 28 ± 2°C. During the period of acclimatization and experimentation the fish were not fed. If the number of deaths exceeded 5% in any batch of fish during acclimatization the concerned batch was discarded. Studies on toxicity were conducted for technical as well as 20% EC, employing static and continuous flow-through systems as recommended in the report of the committee on methods of toxicity tests and aquatic organisms (Anon, 1975). The solutions of desired concentrations were prepared in acetone to yield a concentration of 100 mg/ml. The control fish received an equal quantity of acetone to that of the highest concentration of the toxicant tested.

For flow-through system test, solutions of desired concentrations were prepared once in five hours in glass reservoirs and were let into the test containers through thin-walled polythlene tubes. The flow rate was adjusted with regulators such that three litres of water passed through containers in one hour. The conditions of the test medium were temperature 28 + 2°C, Oxygen 8-10 ppm, Calcium hardness 52 mg/l, alkalinity 472 mg/l and pH 8.1.

Experiments were conducted to determine test concentrations causing mortality of test fish in the range of 10 to 90%. For each concentration, 10 fish were tested and the experiment was repeated thrice. Probit analysis (Finney, 1971) as recommended by Roberts and Boyce (1972) was followed to calculate the LC_{50} values. The technical fenvalerate and 20% EC were supplied by Searle (India) Limited, Bombay.

After their exposure to the pesticide the survived fish tissues were taken for residue analysis. The brain, muscle, stomach, liver, kidney, anterior and posterior parts of the intestine and gill tissues (each 100 mg) were homogenised and centrifuged individually, using acetone as solvent. The extracted solvent was runover through a column, containing activated Florosil as adsorbant. The first three 15 ml aliquots of acetone were pooled and concentrated to 1 ml (Analytical manual EPA, 1974).

The concentrated solvent and technical pesticide were spotted on thin layer chromatographic plate (5 × 20 cm size) with silver nitrate impregnation of silica gel as given in Moats (1966). The spotted plate was developed in a solvent system consisting of benzene and acetone. The developed plate was exposed to UV for 10 minutes and the spots were observed. For residue analysis, the precautions laid down by Thompson (1974) were followed. The entire procedure was standardised for normal fish tissues adding the pesticide. The 100 × Rf values were calculated for different solvent systems and the one that has the best solution was considered as a mobile solvent for TLC.

Observations and Discussion

The LC_{50} values for static and continuous flow-through systems for 48 h and 96 h of two freshwater fishes are given in Table 37.1. The differences between observed and calculated values was tested for significance. The qualitative confirmation of the pesticide in different tissues of fish *L. Rohita* is given in Table 37.2.

The values within the parentheses are confidential limits.

* The values are tested for chi-square and found to be not significant at p 0.05.

The LC_{50} values for fenvalerate to different fishes were given by Chapman *et al*., (1985), Bradbury *et al*., (1984), Mulla *et al*., (1978) and Zitko *et al*., (1977 & 1979). The toxicity was also determined for bobwhite quail (*colinus virginanus*) by Bradbury and Coats (1982), and for cray fish (*Procambarus clarkii*) by Min Long Cheam *et al*., (1980). These authors proved that fenvalerate is highly toxic to fish and formulations are more toxic than technical grade fenvalerate. They also pointed that the LC_{50} value for 96 h is the same as that of 48 h.

In the present study *Labeo rohita* is more sensitive than *Aplocheilus Panchap*. With 20% EC, the range of concentrations producing response of morality for toxic tests is narrow when compared to the technical fenvalerate.

When static and continuous flow through system values of technical fenvalerate are compared in *Labeo rohita*, static value is 2 times higher for 48 h. With 20% EC in both types of tests for the two different fishes the values are not very significant for comparision. The LC_{50} values after 48 h duration are not showing any impact on the test up to 96 h. So the values of LC_{50} for 96 h are similar to the values of 48 h. 20% EC is thousand times more toxic than technical fenvalerate to both the species. The present work can be summarised as follows.

1. Fenvalerate is highly toxic to fish.
2. The static values were higher than continuous flow-through system.
3. 20% EC is more toxic than technical (Coats and O'-Donnel Jeffery, 1979).
4. A synergistic effect can be attributed for high toxicity when formulations are used.
5. The 48th LC_{50} values are the same 96h (Bradbury, 1985).

Table 37.1 : Toxicity of technical Fenvalerate and 20% Emulsifiable concentrate to freshwater fishes, *Labeo rohita* and *Aplocheilus panchax*

	Technical Fenvalerate (ppm)		*20% Emulsifiable*	*Concentration (ppb)*
	48 h	*96 h*	*48 h*	*96 h*
Labeo rohita	*	*	*	*
Static values	10.7	10.65	2.378	2.327
	(11.9-15.82)	(9.583-11.95)	(1.873-3.018)	(2.047-2.643)
Continuous flow values	1.52	1.506	2.02	2.02
	(1.051-2.199)	(1.225-1.852)	(1.895-2.154)	(1.895-2.154)
Aplocheilus pancha	*	*	*	*
Static values	27.01	27.01	16.09	16.09
	(26.16-27.89)	(26.16-27.89)	(15.38-16.83)	(15.38-16.83)
	*	*	*	*
Continuous flow values	14.19	14.19	9.08	9.08
	(12.82-15.69)	(12.82-15.69)	(8.768-9.404)	(8.768-9.404)

Table 37.2 : Residue confirmation of the pesticide in the tissues-Chromatogram

Layer	–	Silica gel
Solvent	–	Benzene + Acetone (85 + 15 v/v)
front	–	12 cm
Spray reagent	–	Silver nitrate
Time	–	50 minutes
UV light	–	10 minutes
Days of exposure	–	96 hours
color of the spot	–	Brown

Tissues	100* Rf
Brain	78.1
Gill	77.7
Liver	77.3
Kidney	79.6
Muscle	79.0
Stomach	Not detectable
Anterior intestine	Not detectable
Posterior intestine	Not detectable

The incorporation of a cyanogroup enhances the stability of the compound in the environment and inclusion of cyano and chloro groups increased the lipohilicity and the potency of toxic action (Casida, 1980 and Elliott, 1976). Apart from the above, the ingredients mixed in the formulation of enhance the toxicity. So, before any legitimate questions are raised regarding environmental policy and planning, the ingradients have to be viewed seriously.

In residue analysis, qualitative confirmation of the residues was observed in brain, gill, liver, kidney and muscle, as the pesticide is having a chloride ion, silver nitrate and impregnation and exposure to UV light produce the spot due to silver and chloride ion binding. The qualitative confirmation locates the site of position, and provides an opportunity to study the nature of metabolites, which are yet to be studied as metabolites are more toxic than parent compound (Tilak *et al.*, 1980, 1981 and Tilak, 1982).

References

Analytical Manual EPA, (1974) Pesticide Analytical Manual, FAO, Washington, D.C.

Anderson, L. Richard., 1982 (Toxicity of Fenvalerate and Permethrin to several non target Invertebrates. Environ, Entomol., II pp. 1251-57.

Anon, (1975) Committee on "Methods of Toxicity tests with fish. Methods of Toxicity tests with Fish, macro-invertebrates and amphibians" EPA Oregon, 61 pp.

Bradbury, S.P. and Coats, J.R., (1982) Toxicity of Fenvalerate to bob white quail (*Colinus Virginianus*) including brain and liver residues associated with mortality. J. Toxicol. Environ. Health, V, 10, pp. 307-19.

Bradbury, S.P., Joel R. Coats, James M. McKim, (1984) Differential toxicity and uptake of two Fenvalerate Formulations in fat head minnows (*Pimphales promelas*). Environmental Toxicology and Chemistry, V. 4, pp. 533-41, USA Pergamon Press Limited.

Casida, J.E., (1980) Pyrethrum flowers and Pyrethroid insecticides, Environ Health Perspect, V. 34, pp. 189-202.

Chapman, G.A., Cueris, L.R. and Sem, W.K., (1985) Toxicity of Fenvalerate to developing Steelhead trout following continuous or intermittant exposure. Journal of Toxicology & Environmental Health, V. 15, pp. 445-57.

Coats, J.R. and Donnell-Jefery. N.L.O., (1979) Toxicity of four synthetic pyrethroid insecticides to rainbow trout. Bull Environ. Contam. Toxicol. V. 23, pp. 250-55.

Elliott, M. (1977) Synthetic Pyrethroids, In Synthetic Pyrethroids M. Elliott, Editor, ACS Aymposium Series 42, Washington, D.C., pp. 1-28.

Finnery, D.J., (1971) Probit Analysis, Cambridge University Press, 333 pp.

Min-Long Cheam, James W. Avault, Jr., and Jerry B. Graves (1980) Acute toxicity of selected Rice Pesticides to Cray Fish, *Procambarus clarkii*. The progress fish culturist, V. 42, No. 3.

Moats, W.A., (1966) Analysis of dairy products for chlorinated insecticide residues by Thin layer Chromatography. J. Assoc. Off. Anal. Chem. V. 49, pp. 795-800.

Mulla, M.A., Nawab-Gojrati, H.A. and Darwazeh, H.A., (1978) Biological activity and longevity of new Synthetic Pyrethroids against mosquitoes and some non-target insects. Mosq. News, V. 38, pp. 90-96.

Rao, D.M.R., Priyamvada Devi, A. and Murty, A.S., (1980) Relative toxicity of endosulfan, its isomers, and formulated products to the freshwater fish, *Labeo Rohita*, J. of Toxicol. and Environ, Health, V. 6, pp. 825-34.

Robert, M. and Boyce, C.B., (1972) Principles of biological assay. In methods in microbiology, eds. J.R. Norris and D.E. Robbins, V. 74, pp. 153-90, New York : Academic.

Southwick, L.J., Smith, S. and Willis, G.H., (1983) Compartmentalization of permethrin on cotton leaves in field during spray application season. Environ, Toxicol. Che. V. 2, pp. 29-34.

Talekar, N.S. (1977) Gas liquid chromatographic determination of - cyano-3-phenoxy benzyl-isopropyl 4-chloro phenylacetate. Residues in cabbage. Journal of the AOAC V. 60, No. 4.

Thompson, J.R., (1975) Analysis of pesticide residues in human and environmental samples.

Tilak, K.S., (1982) Relative toxicity of carbaryl, 1-naphtol and three formulations of carbaryl to *Channa punctata* (Bloch). Matsya, V. 8, pp. 47-48.

Tilak, K.S., Mohan Ranga Rao, D. Priyamvada Devi, A. and Murty, A.S., (1980) Toxicity of carbaryl & 1-naphthol to the freshwater fish *Labeo rohita*. J. Expt. Biol. V. 18, No. 1, pp. 75-76.

Tilak, K.S., Mohan Ranga Rao, D. Priyamvada Devi, A. and Murty, A.S., (1981) Toxicity of carbaryl and 1-naphthol to four species of freshwater fish J. Biosci. V. 3, No. 4, pp. 457-62.

Zitko, V., Carson, W.G. and Metcalfe, C.D., (1977) Toxicity of pyrethroids to Juvenile Atlantic Salmon. Bull Environ Contam Toxicol V. 18, p. 35.

Zitko, V., Leese, D.M.M.C., Metacalfe, C.D. and Carson, W.G., (1979) Toxicity of permethrin and related pyrethroids to salmon and lobster. Bull Environ Contam Toxicol. V. 2, pp. 338-43.

Groundwater Resources of Tambraparni River Basin, Tamil Nadu

A. Balasubramanian
and
J.C.V. Sastri

Introduction

Quantitative and qualitative analysis of the existing water resources of a river basin needs a detailed integrated, hydrogeological approach utilizing the geological, geomorphological, hydrometeorological hydrogeological, hydrochemical, geophysical and water balance methods. An attempt has been made in this work to assess the occurrence, availability and suitability of groundwater resources of Tambraparni river basin, Tamil Nadu.

Tambaraparni river basin, located between latitudes 8°25′ to 9°13′N and longitudes 77°10′ to 78°10′E is lying mainly in the Tirunelveli and partly in the Chidambarnar districts of Tamil Nadu, India (Fig. 38.1). The basin occupies an aerial extent of 5382 sq. km. covering fully the areas of Ambasamudram, Tenkasi, Shencottah, Kadayanallur, Tirunelveli, Srivaikundam taluks and partly the areas of Sankarankovil, Nanguneri, Kayattar and Tiruchendur taluks of the two districts (Survey of India Topographic Sheet Nos. 58 G/4, 8, 12, 16 : 58 H/1, 2, 5, 6, 7, 9, 10, 13, 14 and 58 L/2). The basin is bounded by the Gulf of Manaar in the east and western ghats in the west (Fig. 38.2).

The trunk stream of the basin, Tambraparni, originates at an altitude of 1869 m above MSL, near Agastyamalai in the western ghats and descends to the plains in five beautiful falls at Papanasam and Courtallam. After a run of 120 km. the river confluences with the Gulf of Manaar, 15 km. south of Tuticorin the "Pearl City" of India. Manimuttar, Ghatananadi, Pachayar, Chittar and Uppodai are the major tributaries of the river Tambraparni. About 1347 sq. km. (25%) of the basin area is covered by reserved forests and hills.

Geology

Tambraparni basin is mainly a hardrock terrain, predominantly underlain by crystalline Archaean rocks, which are highly heterogeneous in nature,

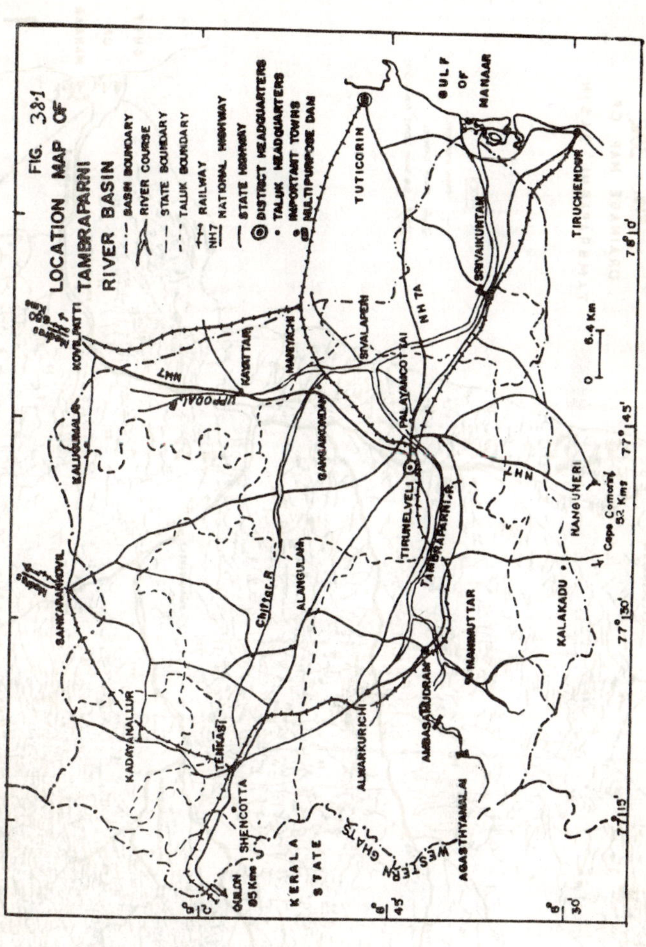
FIG. 38·1
LOCATION MAP OF
TAMBRAPARNI
RIVER BASIN
BASIN BOUNDARY
RIVER COURSE
STATE BOUNDARY
TALUK BOUNDARY
RAILWAY
NH7 NATIONAL HIGHWAY
STATE HIGHWAY
DISTRICT HEADQUARTERS
TALUK HEADQUARTERS
IMPORTANT TOWNS
MULTIPURPOSE DAM
TUTICORIN
GULF OF MANAAR
SRIVAIKUNTAM
TIRUCHENDUR
NH 7A
SIVALAPERI
PALAYAMCOTTAI
TIRUNELVELI
TAMBRAPARNI.R
NANGUNERI
KALAKADU
MANIMUTTAR
AMBASAMUDRAM
AGASTHYAMALAI
ALWARKURICHI
KERALA STATE
WESTERN GHATS
SHENCOTTA
QUILON 85 Kms
TENKASI
KADAYANALLUR
SANKARANKOVIL
KALUGUMALAI
KOVILPATTI
NH7
UPPODAI.R
KAYATTAR
GANGAIKONDAN
ALANGULAM
Chittar.R
0 6.4 Km
77°15'
77°30'
77°45'
78°0'
9°0'
8°45'
8°30'

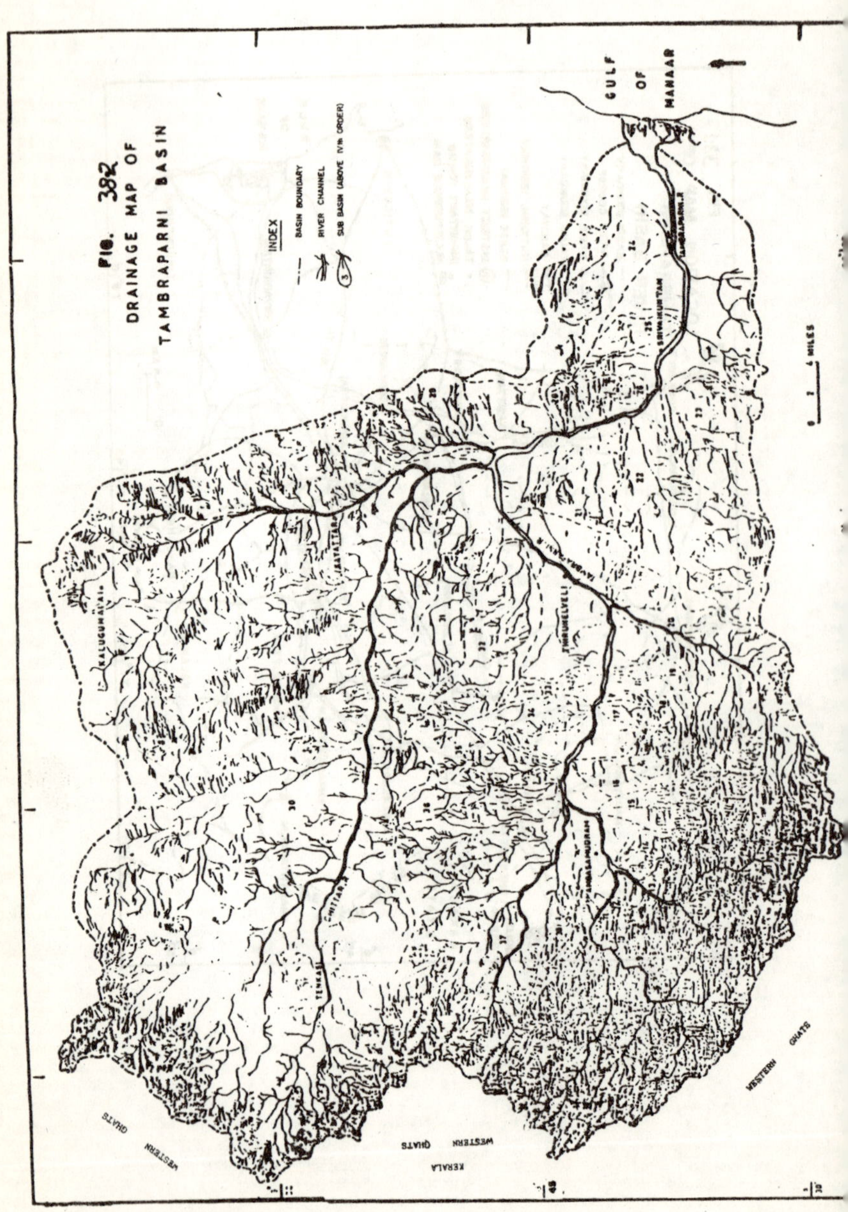
FIG. 382
DRAINAGE MAP OF
TAMBRAPARNI BASIN
INDEX
BASIN BOUNDARY
RIVER CHANNEL
SUB BASIN (ABOVE IVth ORDER)
GULF
OF
MANAAR
0 2 4 MILES
WESTERN GHATS
KERALA
WESTERN GHATS
WESTERN GHATS

posing problems for groundwater occurrence and development. Fig. 38.3 shows the different lithological units encountered in this basin. The stratigraphic sequence of the geological formations occurring are :

Age		*Lithology*
Recent to subrecent	–	Soils, Coastal sands, teri sands, kankar, tufa and laterites, calcareous sandstones and shell-limestones.
Tertiary	–	Hard and compact calcareous sandstones and shell limestones.
Archaean	–	Younger pegmatitic and granitic intrusions.
	–	Composite gneiss, granite-mica-gneiss.
	–	Garnetiferous mica-gneiss, Amphibolites, pyroxene granulites, charnockites.
	–	Crystalline limestones and cal-granulites.
	–	Quartzites.

Three major types of soils occur in Tambraparni basin, namely (1) red sandy soil (thickness 1-5m), (2) black cotton soil (thickness 1-3m) and (3) the river alluvium (thickness 7-12 m). Major portion of the basin is covered with red sandy soil and black cotton soil is confined to the NE portion.

The study of structural and tectonic history of the basin area (Narayanaswami and Purnalakshmi, 1967) revealed that there were several episodes of deformation which caused repeated folds, joints and fractures. The available bore hole litholog records reveal that : (*i*) joints are almost common to all rock types, (*ii*) fractures are confined to gneisses and (*iii*) fissures are seen only in charnockites and (*iv*) the depth to the basement topography ranges from 7.0 m to more than 40 m and is increasing towards the coast line. These structures are the pathways of rapid groundwater movement and recharge. These are the favourable zones for groundwater development by open wells or bore wells.

Hydrometeorology

The plain lands of the basin fall under the semi-arid climatic type and the areas over the adjacent to western ghats are of dry to moist sub-humid climatic types (Rammohan, 1984).

In Tambraparni basin, precipitation is the main source of groundwater. The pattern of precipitation is essentially of a tropical monsoon type

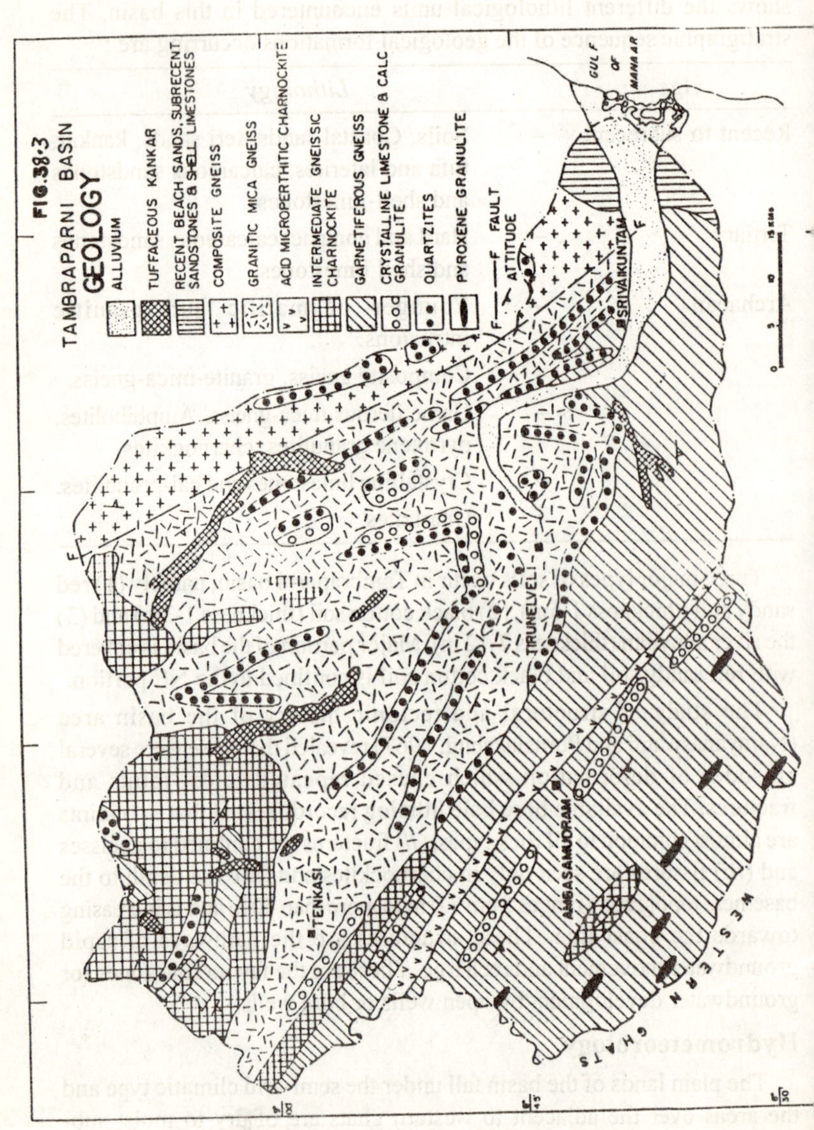
FIG.38·3
TAMBRAPARNI BASIN
GEOLOGY
ALLUVIUM
TUFFACEOUS KANKAR
RECENT BEACH SANDS, SUBRECENT SANDSTONES & SHELL LIMESTONES
COMPOSITE GNEISS
GRANITIC MICA GNEISS
ACID MICROPERTHITIC CHARNOCKITE
INTERMEDIATE GNEISSIC CHARNOCKITE
GARNETIFEROUS GNEISS
CRYSTALLINE LIMESTONE & CALC GRANULITE
QUARTZITES
PYROXENE GRANULITE
F——F FAULT
ATTITUDE
GULF OF MANNAR
SRIVAIKUNTAM
TIRUNELVELI
AMBASAMUDRAM
TENKASI
WESTERN GHATS

where the effect of the winter monsoon is dominant. The average annual isohyets and monthly average bar charts of 12 rain gauge stations of the basin are shown in Fig. 38.4. The average annual precipitation is 88 cm. Table 38.1 shows the climatological and stream flow statistics based on the long term records. The trunk stream is found to be perennial as per the flow records analysed. The actual annual evapotranspiration is found to be 636 mm. The average annual discharge registered by CWC at Murappanadu gauging station is about 418 mcum.

Hydrogeomorphology

Quantitative morphometric analyses have been carried out using 1 inch 1 mile topo-sheets. Following the channel ordering system of Strahler (1957), the basin has been divided into 44 sub-basins (26 of iv; 10 of v; 6 of vi; 1 of vii and 1 of viii orders). The trunk stream has been found to be of IXth order.

Drainage texture is fine to moderate in the hilly terrains and very coarse in plains and coastal zones. The drainage density of sub-basins range from 7 to 1 km/km^2 and it has been found that around 16 sub-basins (19, 21-26, 28-36) bear a very low drainage density (<2.0) indicating the possibility of high groundwater potentiality. Stream frequency is maximum in ghats (24 to 5) than in the plains (4.0 to 9.0) indicative of greater run off and lesser infiltration in the ghats.

Based on terrain analysis using LANDSAT imagery and through visual interpretations it is possible to delineate the hydrogemorpic units for groundwater studies on a regional scale (Thillaigovindarajan, 1972, Roy, 1980; Perumal *et al.*, 1983).

The hydrogemorphic units favourable for tapping groundwater have been demarcated using 1:250000 scale LANDSAT imageries. The groundwater potential zones of the basin derived out of this study have been shown in Fig. 38.5 and other characters of Table 38.2.

Hydrogeology

All types of groundwater resources evaluation necessarily need the study of the occurrence, behaviour of groundwater system and the evaluation of aquifer parameters. Groundwater, in this basin, occurs : (*i*) under water table conditions in the weathered crystalline complex and (*ii*) under semi-unconfined conditions in the coastal tract. The behaviour of groundwater system can be analysed by preparing water table contour maps.

Grid deviation methods of representing the geological data (Saha *et al.*, 1963, Biswas *et al.*, 1967) has been adopted in this study. An average

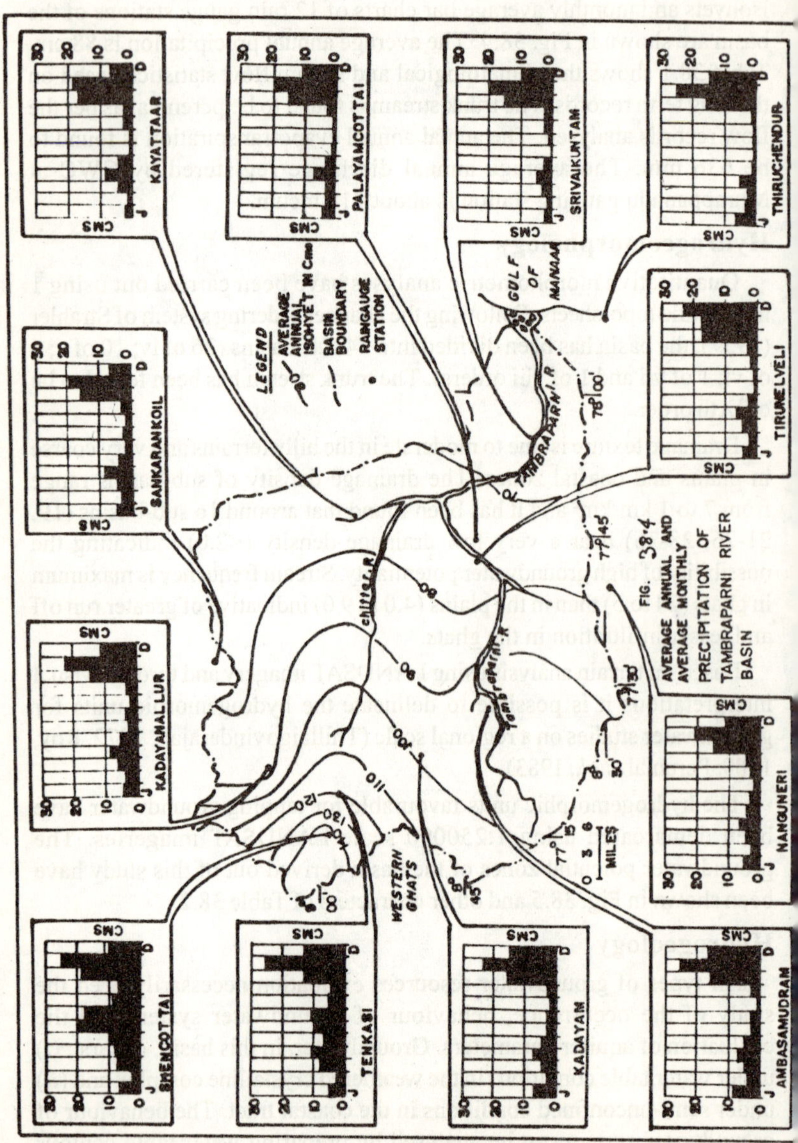

FIG. 38·4 AVERAGE ANNUAL AND AVERAGE MONTHLY PRECIPITATION OF TAMBRAPARNI RIVER BASIN

Table 38.1 : Tambraparni Basin–Climatological Statistics

Factor		*Jan.*	*Feb.*	*Mar.*	*Apr.*	*May*	*June*	*July*	*Aug.*	*Sept.*	*Oct.*	*Nov.*	*Dec.*	*Annual*
+ Potential evaporation transpiration (mm) (Normal)		130.6	136.1	170.6	155.0	175.0	175.5	186.1	194.3	177.4	144.1	107.9	121.2	1873.9
* Actual evaporation transpiration (mm) (Average)		31.6	48.5	57.1	55.3	61.8	42.9	104.7	113.5	100.2	28.2	20.8	26.0	690.0
+	Max	25.1	25.8	27.9	30.1	31.8	30.1	29.9	29.5	28.1	28.4	26.3	25.6	
Temperature (°C)	Min	18.2	18.9	21.1	23.6	24.9	25.0	24.8	24.6	23.8	23.2	21.8	19.6	
	Mean	21.7	22.4	24.5	26.9	28.4	27.6	27.4	27.0	25.9	25.8	24.1	22.6	25.4
* Relative	High	95.0	82.5	87.0	86.5	87.0	85.5	82.5	77.5	91.0	95.5	95.5	93.0	
Humidity (%)	Low	33.0	27.0	43.0	52.5	45.0	54.5	51.5	40.0	36.0	47.5	45.0	45.0	
	Mean	64.0	54.8	65.0	69.5	66.0	70.0	67.0	58.8	63.5	71.5	70.0	69.0	65.8
	Max	7.8	9.1	10.6	10.1	12.1	13.8	14.0	14.0	12.8	10.6	7.6	6.9	
* Evaporation	Min	4.4	4.1	5.0	3.3	3.5	6.8	7.0	7.8	4.1	2.5	0.9	2.2	
(mm)	Mean	6.1	6.6	7.8	6.7	7.8	10.3	10.5	10.9	8.4	6.6	4.3	4.6	90.6
* Soil Moisture content (%)		20.2	18.6	17.8	16.7	15.5	15.2	12.3	10.1	12.2	14.5	27.4	29.8	17.5
* Mean wind velocity (Kmph)		8.2	7.3	6.7	8.6	13.7	30.6	22.7	23.2	15.0	8.3	7.1	6.5	13.2
Average Discharge at Murappanadu (Mm3)		23.81	26.2	14.8	22.9	19.1	23.5	20.7	15.4	16.6	31.0	367.8	196.8	418.0

+ - Meteorological observatory at Palayamcottai; (*) P.W.D. Watershed observatory at Malaipatti.

Table 38.2 : Hydrogeomorphic Units and Groundwater Potential Zones of Tamuraparni Basin

Groundwater Potential	*Surface Drainage*	*Lineament Density*	*Lithology units*	*Major Geomorphic units*	*Suitability units*
Very High	Well to imperfect	Quite long lineament associated with stream channels	Unconsolidated sand, silt and gravel	Older/New flood plains of large stream, Burried Valleys	For intensive groundwater development
High	Parallel, dendritic and anastomatic	High	Granitoid gneiss, Colluviofluvial	Piedmont zone (Bazada) Buried pediments	Suitable for Dug to bore wells
Moderate	Moderately drained	Long to Moderate density	Granitoid gneiss	Flood plain of small streams, Exhumed Surfaces	Dug well of large diameter
Low to Moderate	Sub-dendritic	High density	Granitoid gneiss charnockite	Fresh pediment zone, peripediment zone	For dug wells
Low	–	High density	Granitoid gneiss	Residual hills	–

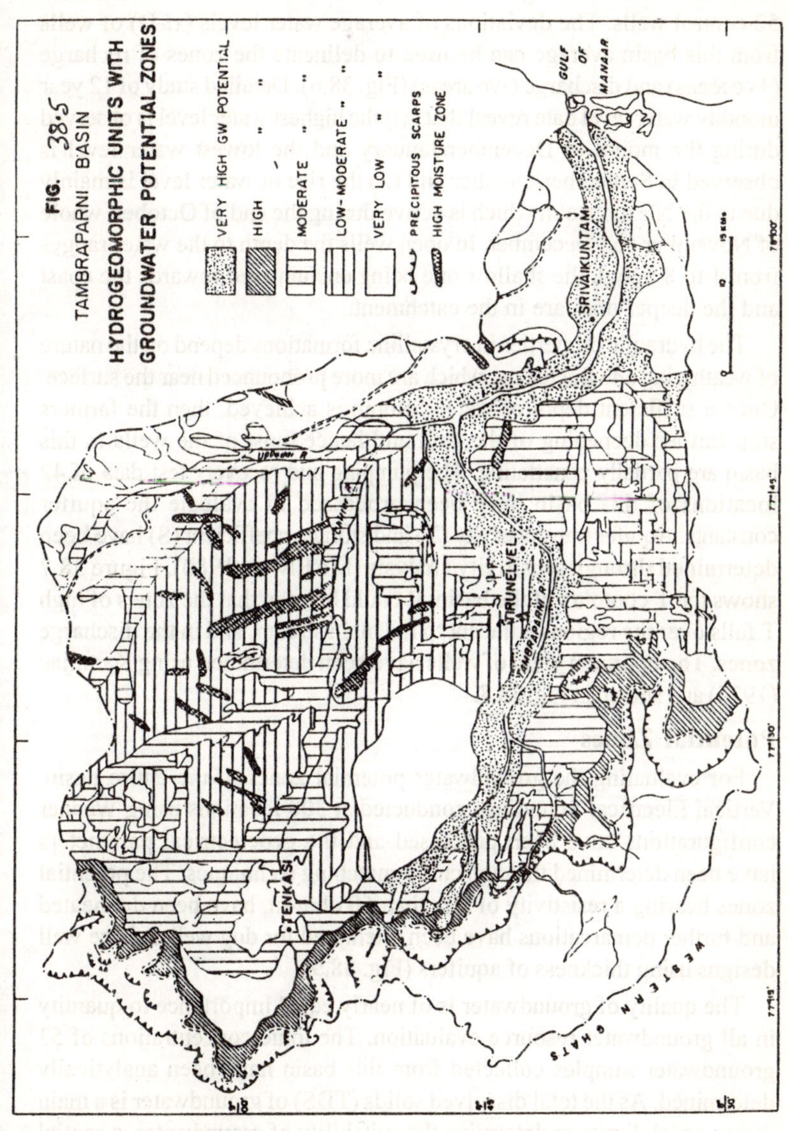
FIG. 38.5
TAMBRAPARNI BASIN
HYDROGEOMORPHIC UNITS WITH
GROUNDWATER POTENTIAL ZONES
VERY HIGH GW POTENTIAL
HIGH " "
MODERATE " "
LOW- MODERATE " "
VERY LOW " "
PRECIPITOUS SCARPS
HIGH MOISTURE ZONE
GULF OF MANAAR
SRIVAIKUNTAM
TIRUNELVELI
TAMBRAPARNI R.
TENKASI
WESTERN GHATS
15 km
77°15'
77°30'
77°45'
78°00'

water level (*xi*) for each control well : (*i*) has been computed using long term data. A basin average water level (D) is determined for the available 52 control wells. The deviations of average water levels (*xi*-D) of wells from this basin average can be used to delineate the zones of recharge (+ve areas) and discharge (-ve areas) (Fig. 38.6). Detailed study of 12 year monthly water level data reveal that : (*i*) the highest water level is observed during the month of December/January and the lowest water level is observed in September/October and (*ii*) the rise of water level is mainly due to the NE monsoon which is active during the end of October, whole of November and December. In open wells the depth to the water ranges from 1 to 8 m bgl, the shallow one being encountered towards the coast and the deeper ones are in the catchment.

The hydraulic properties in crystalline formations depend on the nature of weathering and fracturing, which are more pronounced near the surface. Once a sufficient depth of water column is achieved, then the farmers stop further deepening of the well and hence most of the wells in this basin are partially penetrating. The pumping and recovery test data of 42 locations of this basin have been processed to evaluate the aquifer constants. Aquifer transmissivity (T) and storage coefficient (S) have been determined through sensitivity analysis (Mc Elwee, 1980). Figure 38.7 shows the T contours of the basin. It could be seen that the zones of high T falls over the regions adjacent to the main stream, and in the discharge zones. The optimum yield of wells have been determined using Karanjac (1975) and shown in Fig. 38.8.

Potential Zones

For evaluating the groundwater potential zones of the entire basin, Vertical Electrical Soundings conducted at 300 locations using Wenner configurations have been processed and the geoelectrical parameters have been determined through curve-matching techniques. The potential zones bearing a resistivity of less than 60 ohm-m, have been delineated and further demarcations have been attempted for dug well or bore well designs using thickness of aquifers (Fig. 38.9).

The quality of groundwater is of nearly equal importance to quantity in all groundwater resource evaluation. The ionic concentrations of 53 groundwater samples collected from this basin have been analytically determined. As the total dissolved solids (TDS) of groundwater is a main factor which limits or determine the suitability of groundwater, a spatial variation map of this parameter has been prepared (Fig. 38.10) and the potable zones demarcated (TDS < 500 ppm suitable for domestic use; 500-1000 for agricultural and industrial use; and > 1000 only for industrial use).

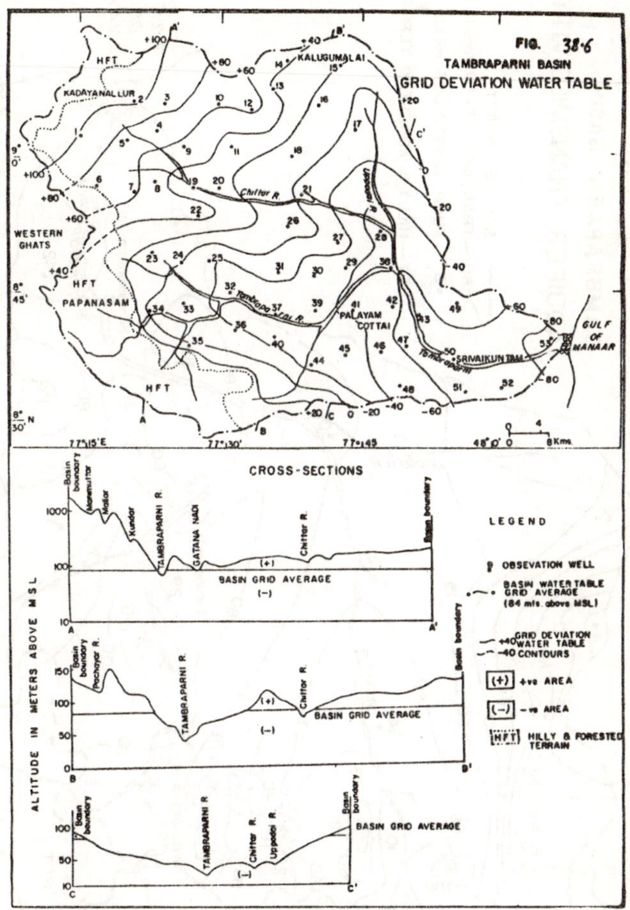
FIG. 38.6
TAMBRAPARNI BASIN
GRID DEVIATION WATER TABLE
CROSS-SECTIONS
LEGEND
OBSEVATION WELL
BASIN WATER TABLE GRID AVERAGE (84 mts. above MSL)
GRID DEVIATION WATER TABLE CONTOURS
(+) +ve AREA
(–) –ve AREA
HFT HILLY & FORESTED TERRAIN
ALTITUDE IN METERS ABOVE MSL
BASIN GRID AVERAGE
WESTERN GHATS
PAPANASAM
KADAYANALLUR
KALUGUMALAI
PALAYAM COTTAI
SRIVAIKUNTAM
GULF OF MANAAR
Chittar R.
TAMBRAPARNI R.

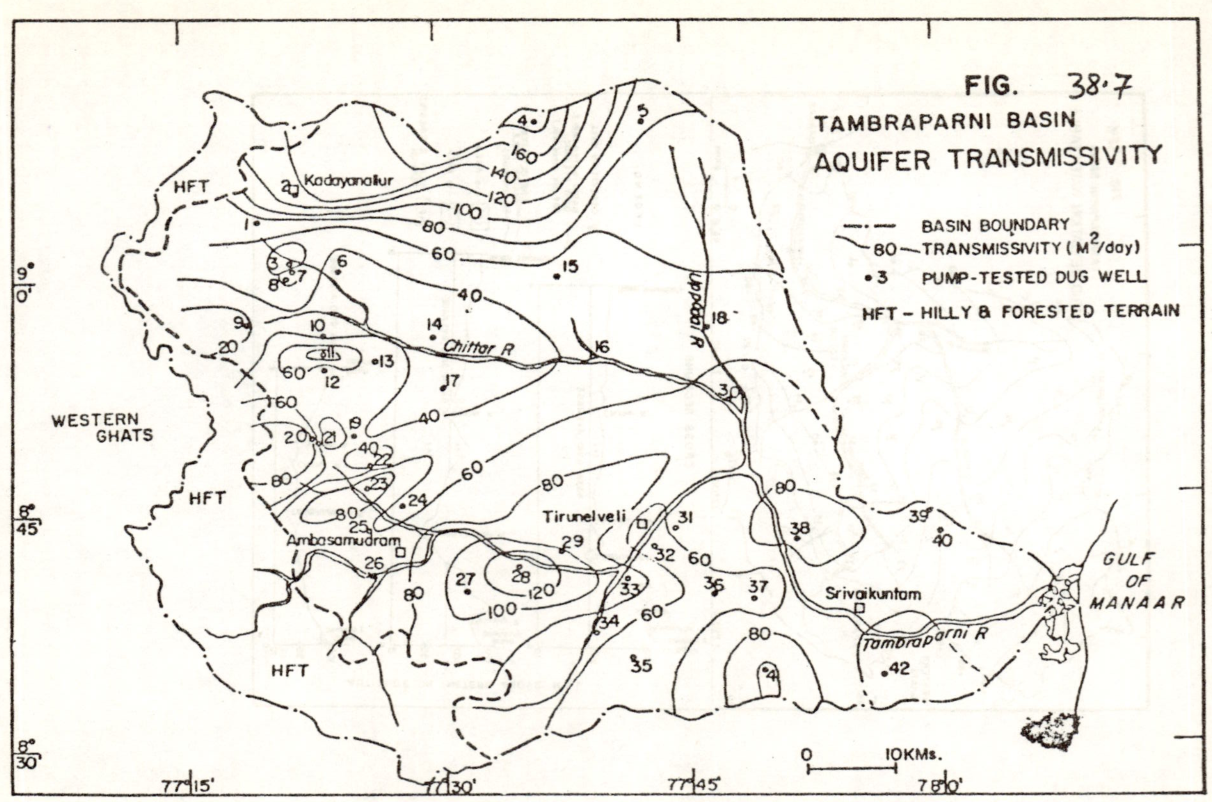
FIG. 38.7
TAMBRAPARNI BASIN
AQUIFER TRANSMISSIVITY
BASIN BOUNDARY
80 — TRANSMISSIVITY (M²/day)
•3 PUMP-TESTED DUG WELL
HFT - HILLY & FORESTED TERRAIN
Kadayanallur
Chittar R
Uppodai R
WESTERN GHATS
HFT
Tirunelveli
Ambasamudram
Srivaikuntam
Tambraparni R
GULF OF MANAAR
0 10KMs.
9° 0'
8° 45'
8° 30'
77°15'
77°30'
77°45'
78°0'

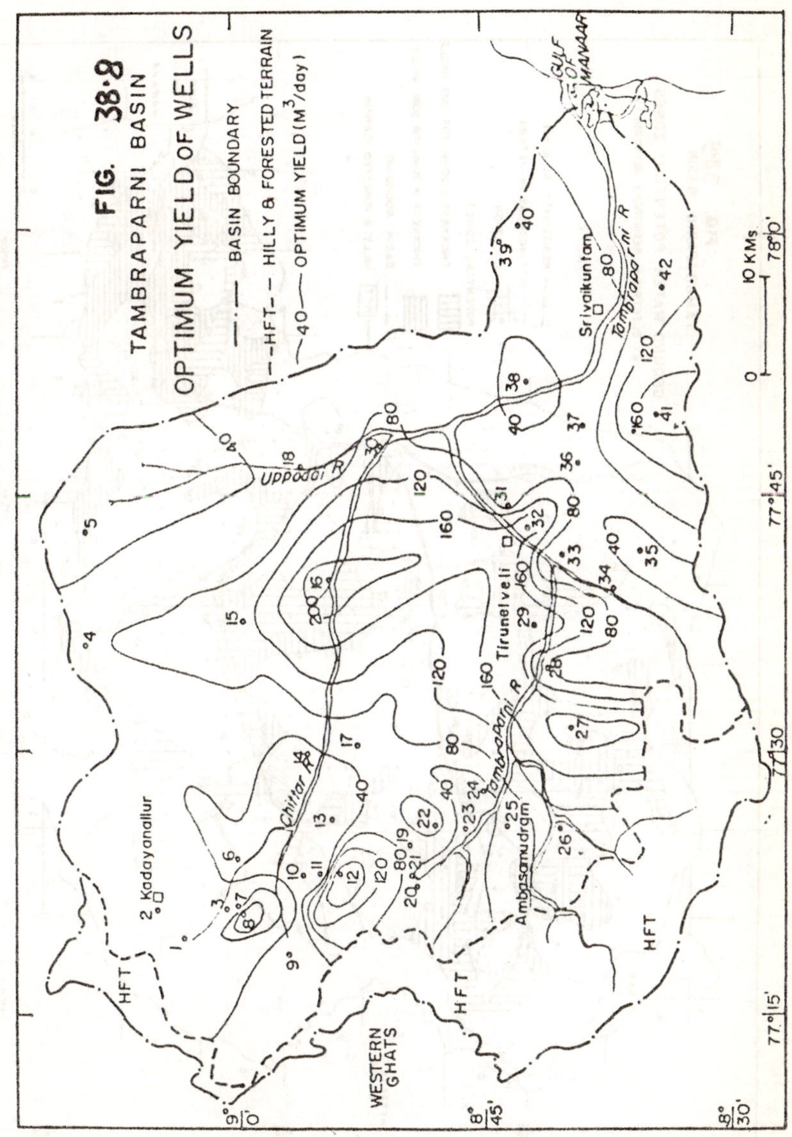
FIG. 38·8
TAMBRAPARNI BASIN
OPTIMUM YIELD OF WELLS
BASIN BOUNDARY
HFT HILLY & FORESTED TERRAIN
40 OPTIMUM YIELD (M³/day)
WESTERN GHATS
Kadayanallur
Chittar R
Uppodai R
Tambraparni R
Tirunelveli
Ambasamudram
Srivaikuntam
GULF OF MANAAR
10 KMs
9°0'
8°45'
8°30'
77°15'
77°30
77°45'
78°0'

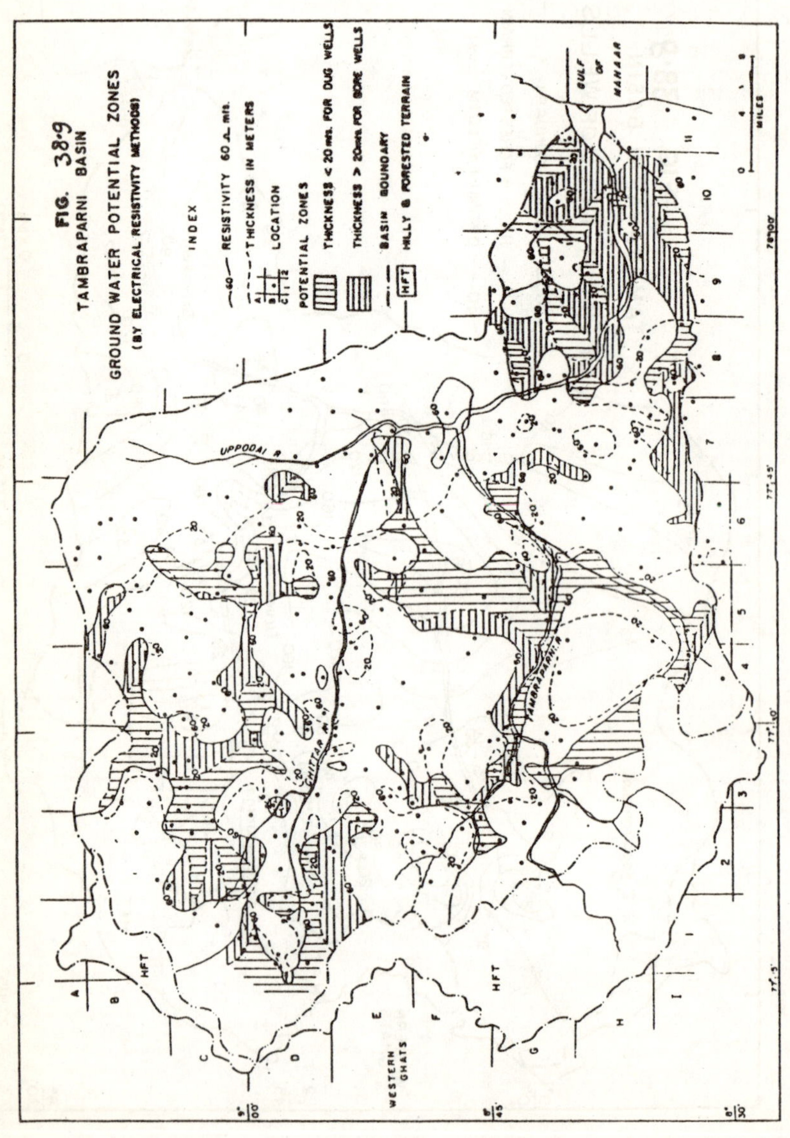
FIG. 38·9
TAMBRAPARNI BASIN
GROUND WATER POTENTIAL ZONES
(BY ELECTRICAL RESISTIVITY METHODS)
INDEX
RESISTIVITY 60 Ω mts.
THICKNESS IN METERS
LOCATION
POTENTIAL ZONES
THICKNESS < 20 mts. FOR DUG WELLS
THICKNESS > 20mts. FOR BORE WELLS
BASIN BOUNDARY
HFT HILLY & FORESTED TERRAIN
UPPODAI R
CHITTAR R
TAMBRAPARNI R
GULF OF MANAAR
WESTERN GHATS
HFT
MILES

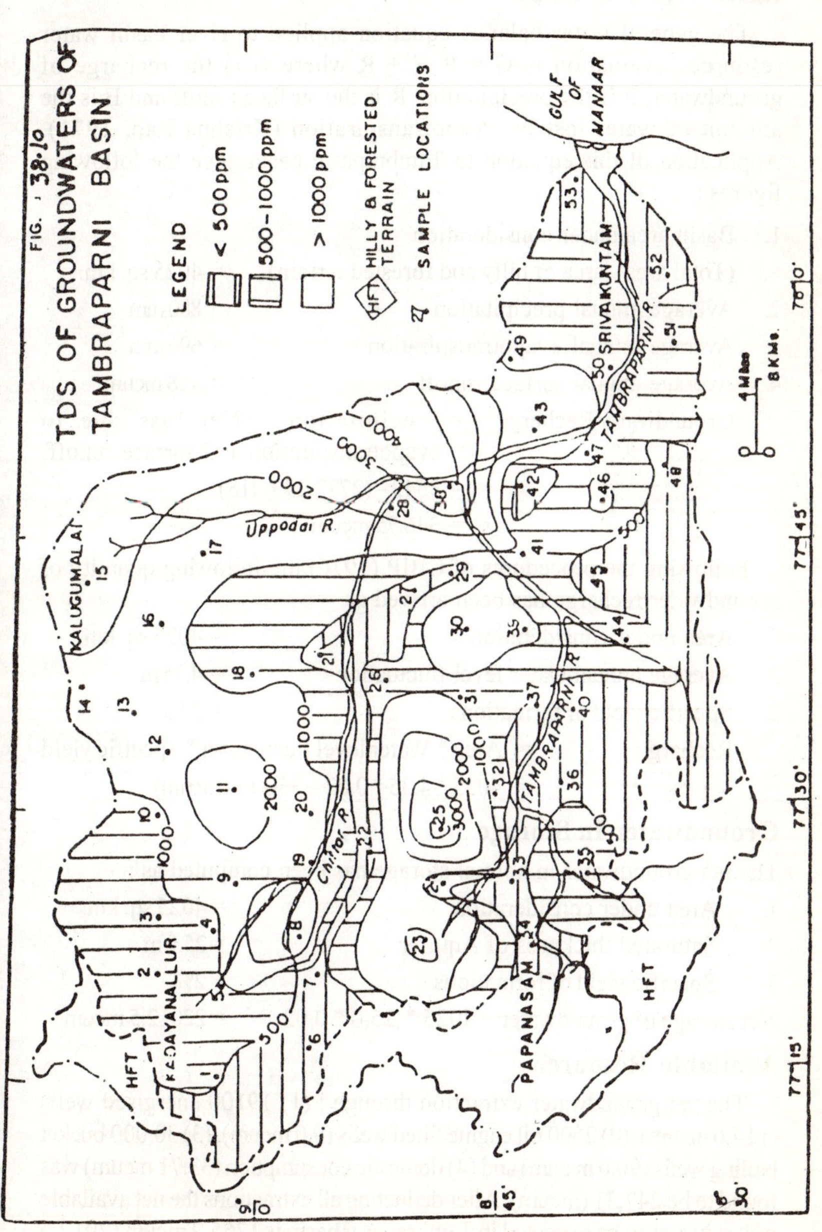
FIG. 38·10
TDS OF GROUNDWATERS OF
TAMBRAPARNI BASIN
LEGEND
< 500 ppm
500-1000 ppm
> 1000 ppm
HFT HILLY & FORESTED TERRAIN
SAMPLE LOCATIONS
GULF OF MANAAR
KALUGUMALAI
KADAYANALLUR
PAPANASAM
SRIVAKUNTAM
Uppodai R.
Chittar R.
TAMBRAPARNI R.
HFT
4 Miles
8 KMs.
77°15'
77°30'
77°45'
78°0'
9°0'
8°45'
8°30'

Basin Water-Balance

The general water-balance equation applied to river basin water resources evaluation is G = P–E + R where G is the recharge of groundwater, P is the precipitation, R is the surface runoff and E is the amount of water lost by evapotranspiration (Krishna Rao, 1971). Application of this equation to Tambraparni basin gave the following figures :

1. Basin area under consideration (Total area - area of hilly and forested terrain) = 4025 sq. km.
2. Average annual precipitation = 880 mm
3. Average annual evapotranspiration = 690 mm
4. Average annual surface runoff = 418 mcum

Groundwater Recharge = Precipitation – Net loss due to evapotranspiration and surface runoff
= 3542 – (2777.25 + 418)
= 346.75 mcum

Following the procedures of CBIP (1976) the following quantity of groundwater recharge has been arrived at

1. Area under consideration = 4025 sq. km.
2. Average annual water level fluctuation = 4.35 m
3. Specific yield of formations = 2 %

Recharge = Area * Water level fluctuation * Specific yield
= 4025 * 4.35 * 0.02 = 350.18 (mcum)

Groundwater In Storage

The net groundwater in normal storage has been computed as :

1. Area under consideration = 4025 sq. km.
2. Saturated thickness of Aquifer = 25.0 m
3. Specific yield of formations = 2%

Net storage of groundwater = 4025 * 25.0 * 0.02 = 22012.5 mcum

Available Resource

The net groundwater extraction through : (1) 19100 energised wells (114.6 mcum), (2) 2500 oil engine fitted wells (9.0 mcum), (3) 30,000 bucket bailing wells (90.0 m cum) and (4) domestic consumption (33.71 mcum) was found to be 247.31 (mcum). After deducting all extractions the net available groundwater to be expected in Tambraparni basin is 1765.2 mcum (2012.5 - 247.3 = 1765.2 mcum)

References

Biswas, A.B. and Chatterjee, P.K., (1967) On representation of water table by grid-deviation method. Bull. Geol. Soc. Ind. V. 4, pp. 12-14.

Karnjac, J., (1975) Brief note on testing the possibility of defining optimum yield of dugwells of larger diameter, Groundwater, V. 13, pp. 4.

Krishna Rao, P.R., (1971) Hydrometeorological aspects of estimating groundwater potential, Sem. Vol. Groundwater potential in hard rock areas of India, Bangalore, pp. 1-2.

Mc Elwee, C.D., (1980) The Thesis equation; evaluation, sensitivity to storage and transmissivity and automated fit of pump test data, Kansas Geol. Surv. Groundwater Ser-3. p. 49.

Narayanaswami, S. and Poorna Lakshmi (1967) Charnockite rocks of Tirnelvelly District, Madras, J. Goel. Soc. Ind. V. 8, pp. 38-50.

Perumal, A. and Roy, A.K. (1983) Application of LANDSAT and aerial data to delinate the hydromorphogeologic zones in parts of Vaigai, Manimuttar and Pambar river basins. Tamil Nadu State, Proc. Nat. Symp. Remote sensing in Developt. and Manag. Wat. Res., New Delhi, pp. 315-19.

Ram Mohan, H.S., (1984) A climatological assessment of the water resources of Tamil Nadu. Ind. Jour. Power and River Valley Developt., pp. 58-63.

Roy, A.K., (1980) LANDSAT (MSE) imagery interpretation of geochemical data from the Sargipalli lead-zinc mine area, Sundergarh district, Orissa (India), Jour. Geochem. Explo., V. 14, pp. 245-64.

Saha, A.K. and Chakravarthy, S.K., (1963) A quantitative petrological study of the Bahalda road Granodiorite, Mayurbhunj, Jour. Deccan Geol. Soc., 3, pp. 50-59.

Strahler, A.N., (1957) Quantitative analysis applied to fluvially eroded land forms. Bull. Geol. Soc. Am., V. 69, pp. 279-300.

Thillaigovindarajan, S., (1972) Geohydrologic survey by aerial photointerpretation. Proc. Sem. On problems in Groundwater Developt., p. 171.

39

Groundwater Potential Zones in the Coastal Aquifers of Tuticorin, Tamil Nadu

V. Radhakrishnan,
R. Thirugnana Sambandam,
R. Chellasamy
and
A. Balasubramanian

Introduction

Though coastal aquifers contain much needed groundwater resources, their study, conservation and exploitation must take into account the presence of sea water and the mixture of fresh and sea waters. Although the general principles are well known, the presence of fluid of variable density and the fact that a small percentage of salt water can make water unusable and complicates the problem. It is aggravated normally by the great influence of local aquifer characteristics and heterogeneties. Salt water intrusion is a major hazard to the public in the coastal aquifers of India. The existing technology is inadequate for the exploitation and management of fresh water aquifers in a saline environment. Better water management practices to limit the encroachment and to tap fresh water resources could be proposed only through an integrated hydrogeological approach. Such an attempt is made to evaluate the groundwater resources of the coastal aquifers of Tuticorin, Tamil Nadu.

Tuticorin coast (Fig. 39.1), which encompasses an International Seaport, is located just a few kilometres north of Tambraparni delta and the pilgrimage-cum-tourist resort Tiruchendur. The study area has been limited to latitudes 8°15′ to 9°0′N and longitude 77°50′ to 78°15′E with an areal extent of 2000 sq. km. It is entirely located in the Chidambaranar district, Tamil Nadu covering major parts of Tuticorin, Ottapidaram, Srivaikundam, Tiruchendur and Sathankulam taluks (Survey of India topographic sheets 58L/1, 2, 3, 4, 5; H/14 and 15). The increasing number of settlements, saltpans and salt based industries call for an increased development of the available water resources.

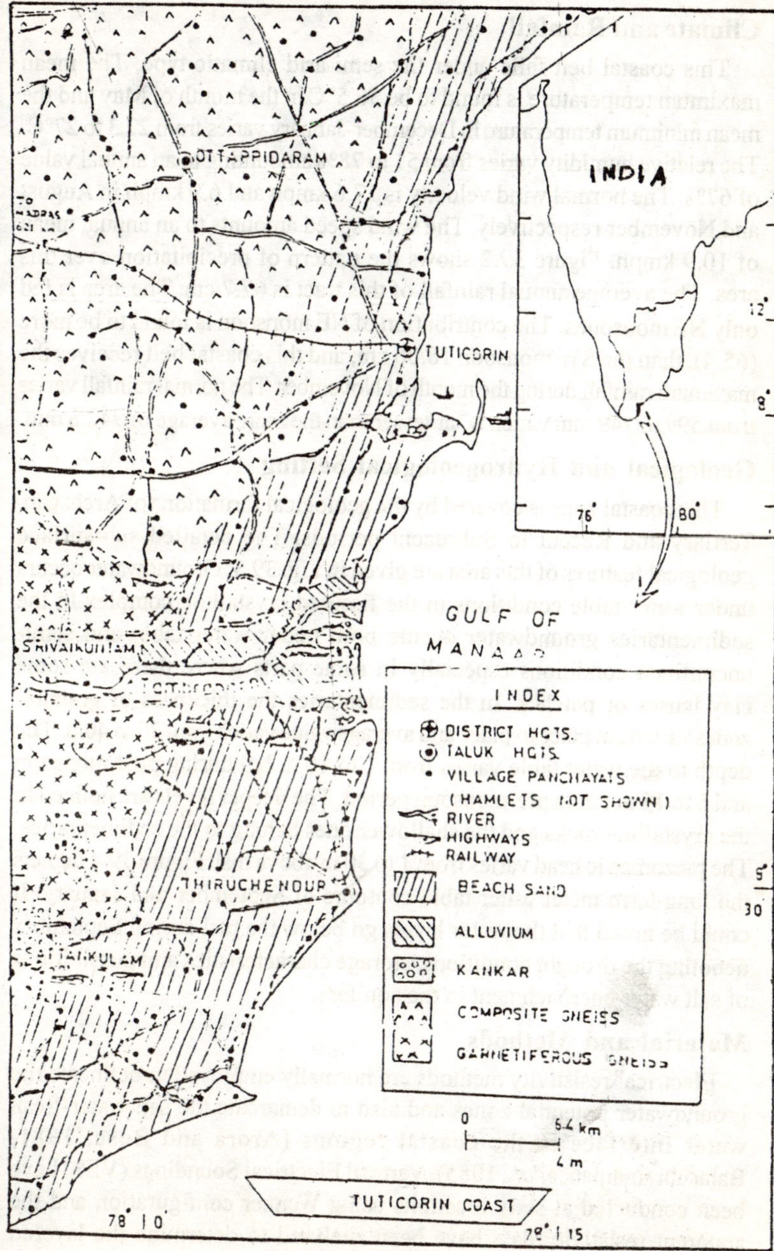
OTTAPPIDARAM
INDIA
TO MADRAS
TUTICORIN
12
8
76
80
GULF OF
SRIVAIKUNTAM
THIRUCHENDUR
SATTANKULAM
INDEX
DISTRICT HQTS.
TALUK HQTS.
VILLAGE PANCHAYATS
(HAMLETS NOT SHOWN)
RIVER
HIGHWAYS
RAILWAY
BEACH SAND
ALLUVIUM
KANKAR
COMPOSITE GNEISS
GARNETIFEROUS GNEISS
0
5.4 km
TUTICORIN COAST
78 10
78 15

Fig. 39.1

Climate and Rainfall

This coastal belt falls under the semi-arid climatic type. The mean maximum temperature is found to be 36.5°C in the month of May and the mean minimum temperature in December-January varies from 22.3 to 27°C. The relative humidity varies from 51 to 78% amounting to an annual value of 67%. The normal wind velocity is 17.6 kmph and 6.9 kmph in August and November respectively. The wind speed amounts to an annual mean of 10.9 kmph. Figure 39.2 shows the pattern of precipitation over this area. The average annual rainfall of this tract is 67.7 cm. The area is fed only NE monsoons. The contribution of NE monsoon is found to be more (65.4), than the SW monsoon 18.06 cm, and this coastal belt receives the maximum rainfall during the month of November. The normal rainfall varies from 599 to 749 mm which is far lesser than the State average of 942.8 mm.

Geological and Hydrogeological Setting

This coastal zone is covered by the geological formations of Archaean, Tertiary and Recent to Subrecent periods. The detailed sub-surface geological features of this area are given in Fig. 39.3. Groundwater occurs under water table conditions in the fissured crystalline complex.In the sedimentaries groundwater occurs both under water table and semi-unconfined conditions especially in some parts where there are some clay lenses or patches. In the sedimentaries the thickness of granular zones vary from place to place and average value varies from 20 to 40m. The depth to the water table varies from 1 to 10 m bgl during post-monsoon and 5 to 15 m bgl in pre-monsoon period. The deeper levels are noticed in the crystalline rocks and the shallower ones belong to the sedimentaries. The piezometric head varies from 1 to 30 m above msl. Figure 39.4 depicts the long-term mean water table contours of September and January. It could be noted that the water levels go below the msl during September denoting the drought conditions, storage characteristics and possibilities of salt water encroachment in the aquifers.

Material and Methods

Electrical resistivity methods are normally employed to delineate the groundwater potential zones and also to demarcate the salt water-fresh water interface in the coastal regions (Arora and Bose, 1981, Balasubramanian, *et al.*, 1985). Vertical Electrical Soundings (VES) have been conducted at sixty locations using Wenner configuration and the apparent resistivity data have been analysed to determine the layered geoelectric parameters. The resistivity of first, second and third layers

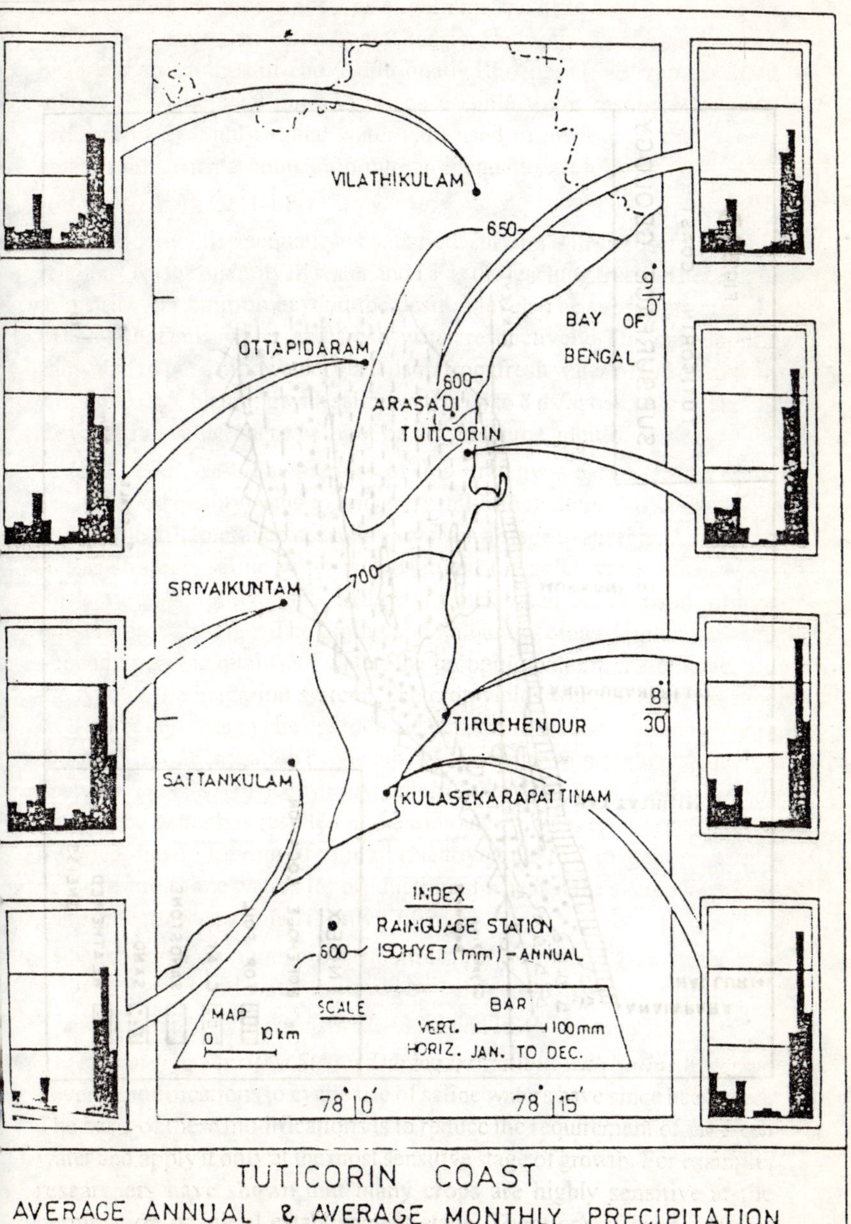
VILATHIKULAM
650
600
OTTAPIDARAM
ARASADI
TUTICORIN
BAY OF
BENGAL
9°
0'
700
SRIVAIKUNTAM
TIRUCHENDUR
8°
30'
SATTANKULAM
KULASEKARAPATTINAM
INDEX
RAINGUAGE STATION
600 ISOHYET (mm) - ANNUAL
MAP SCALE BAR
0 10 km
VERT. 100 mm
HORIZ. JAN. TO DEC.
78°10'
78°15'
TUTICORIN COAST
AVERAGE ANNUAL & AVERAGE MONTHLY PRECIPITATION

Fig. 39.2

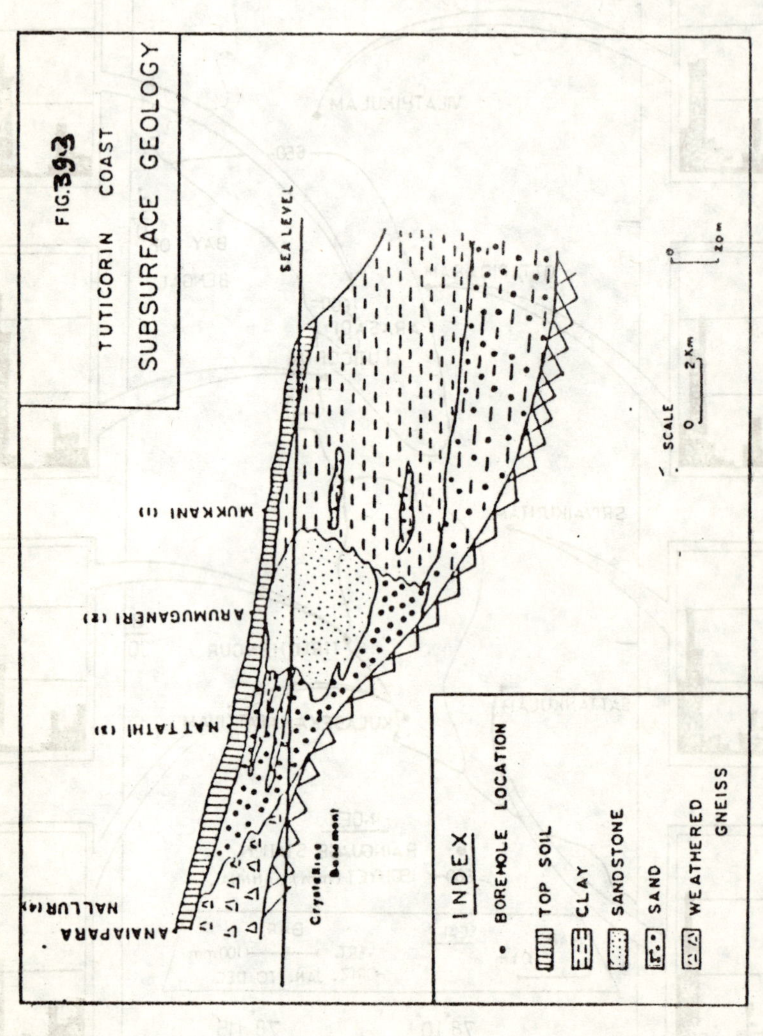
FIG.39.3
TUTICORIN COAST
SUBSURFACE GEOLOGY
SEA LEVEL
MUKKANI (1)
ARUMUGANERI (2)
NATTATHI (3)
ANAIPARA NALLUR (4)
Crystalline Basement
INDEX
BOREHOLE LOCATION
TOP SOIL
CLAY
SANDSTONE
SAND
WEATHERED GNEISS
SCALE
0
2 Km
0
20 m

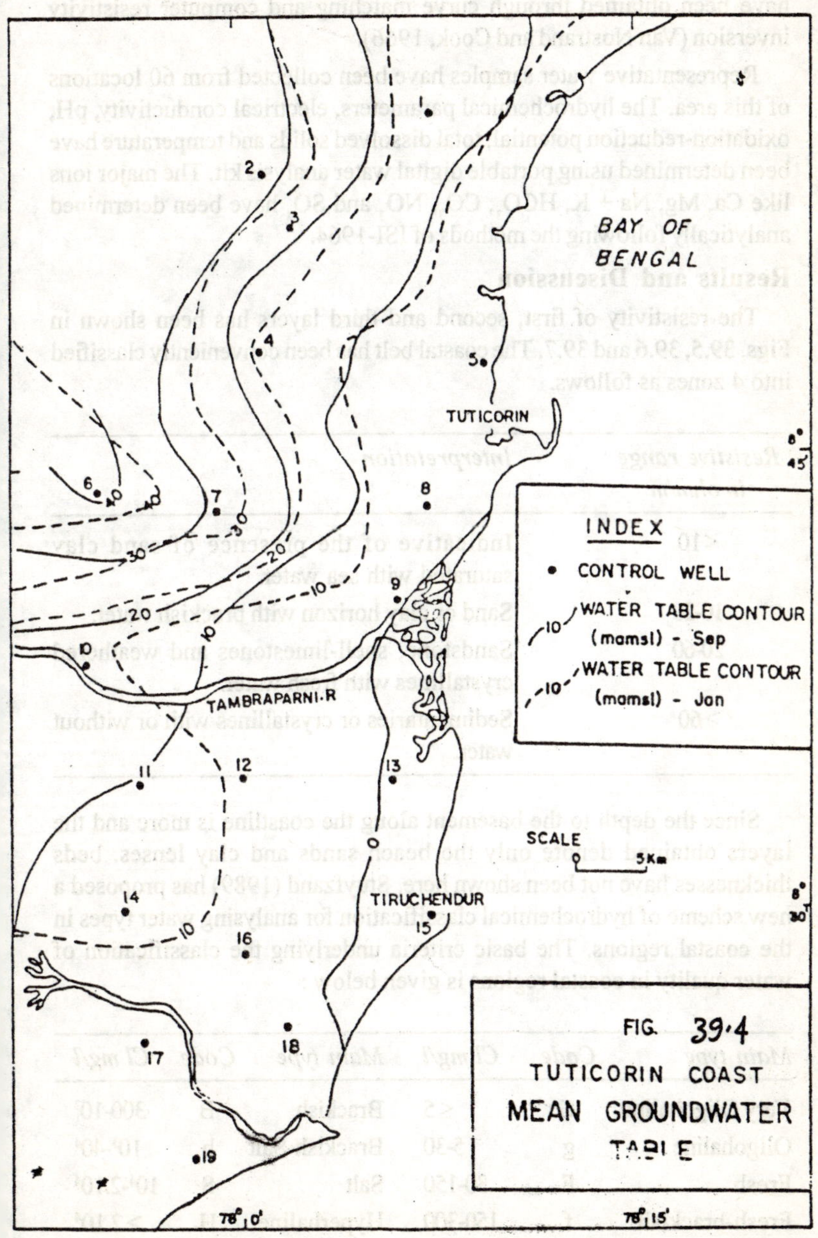

FIG. 39.4
TUTICORIN COAST
MEAN GROUNDWATER
TABLE

have been obtained through curve matching and computer resistivity inversion (Van Nostrand and Cook, 1966).

Representative water samples have been collected from 60 locations of this area. The hydrochemical parameters, electrical conductivity, pH, oxidation-reduction potential, total dissolved solids and temperature have been determined using portable digital water analysis kit. The major ions like Ca, Mg, Na + K, HCO_3, CO_3, NO_3 and SO_4 have been determined analytically following the methods of ISI-1964.

Results and Discussion

The resistivity of first, second and third layers has been shown in Figs. 39.5, 39.6 and 39.7. The coastal belt has been conveniently classified into 4 zones as follows.

Resistive range in ohm/m	*Interpretation*
<10	Indicative of the presence of sand clay saturated with sea water.
10-20	Sand or clay horizon with brackish water.
20-60	Sandstone, shell-limestones and weathered crystallines with fresh water.
>60	Sedimentaries or crystallines with or without water.

Since the depth to the basement along the coastline is more and the layers obtained denote only the beach sands and clay lenses, beds thicknesses have not been shown here. Stuyfzand (1989) has proposed a new scheme of hydrochemical classification for analysing water types in the coastal regions. The basic criteria underlying the classification of water quality in coastal regions is given below :

Main type	*Code*	*Cl mg/l*	*Main type*	*Code*	*Cl mg/l*
Very Oligohaline	G	<5	Brackish	B	$300\text{-}10^3$
Oligohaline	g	5-30	Brackish-Salt	b	$10^3\text{-}40^4$
Fresh	F	30-150	Salt	S	$10^4\text{-}2.10^4$
Fresh-brackish	f	150-300	Hyperhaline	H	$>2.10^4$

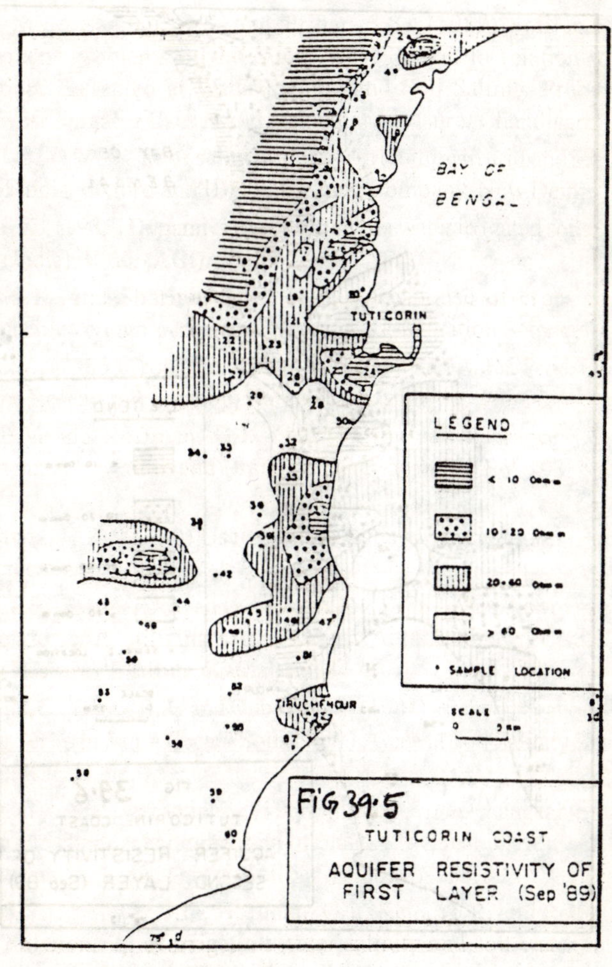

FIG 39·5
TUTICORIN COAST
AQUIFER RESISTIVITY OF FIRST LAYER (Sep '89)

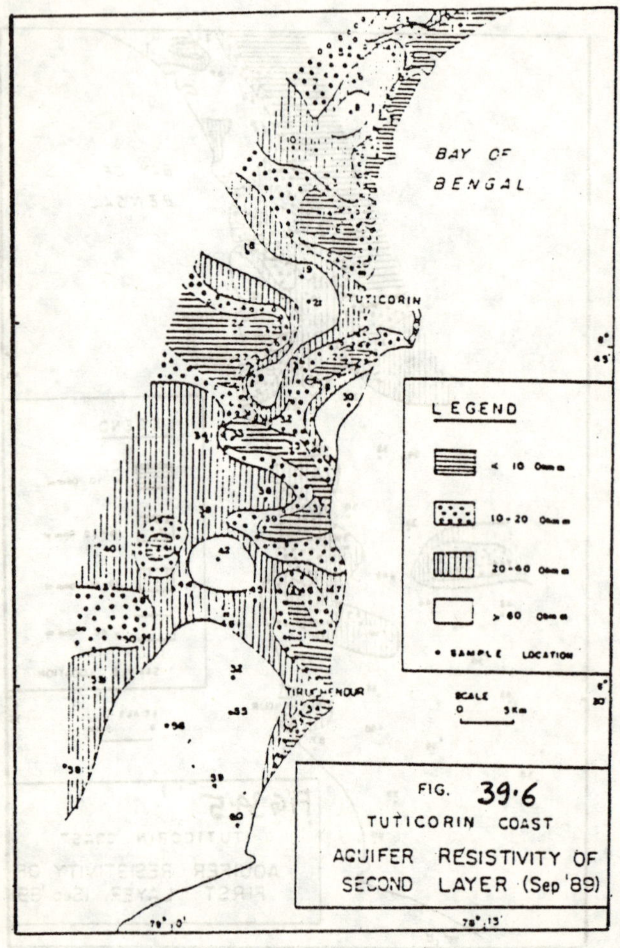

FIG. 39·6
TUTICORIN COAST
AQUIFER RESISTIVITY OF SECOND LAYER (Sep '89)

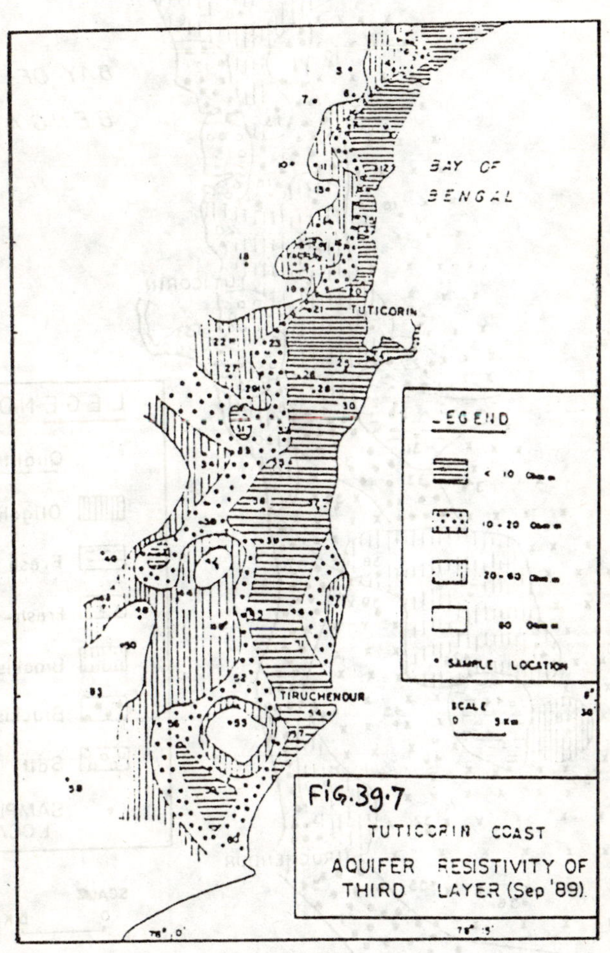

FIG. 39.7
TUTICORIN COAST
AQUIFER RESISTIVITY OF THIRD LAYER (Sep '89).

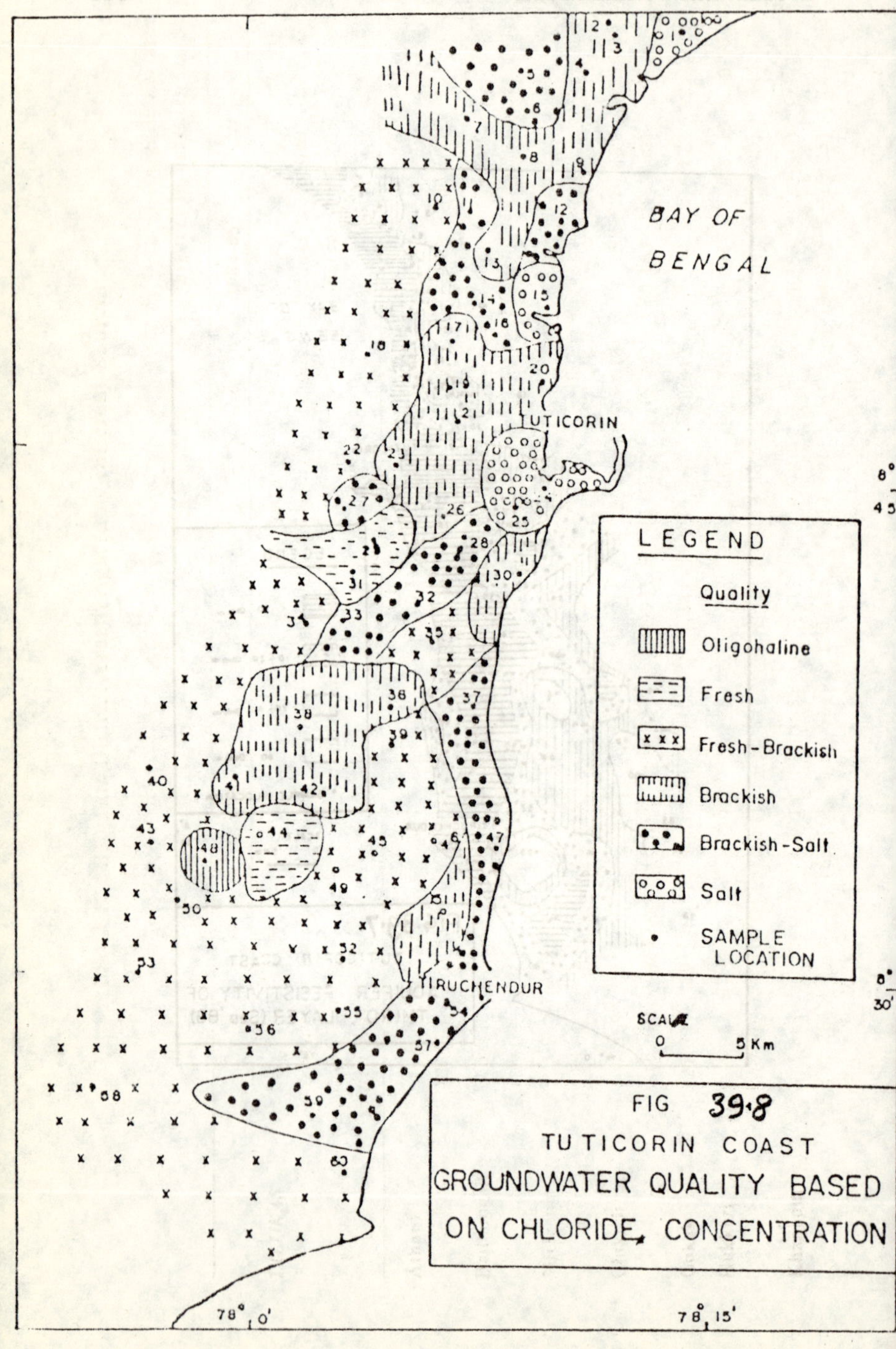

FIG 39·8
TUTICORIN COAST
GROUNDWATER QUALITY BASED ON CHLORIDE CONCENTRATION

Figure 39.8 shows the qualitative characteristics of groundwater existing in this coastal belt based on chloride concentration. The potable zones can be developed within safe yield conditions and keeping in mind of the possibility of saline intrusion.

Conclusion

Spatial distribution of fresh water and saline water along the Tuticorin coast has been studied and presented through hydrogeological, geoelectrical and geochemical studies. Over exploitation of groundwater shall result the mixing of fresh water and saline water which will affect severely both industrial and domestic sectors. In order to protect the potable groundwater zones, safe yield policy is suggested which in turn, leave concentrated salt water zones on which salt-based industries depend.

Acknowledgements

The authors thank the Department of Science and Technology for providing the financial assistance and Prof. C. Pakkiam, Principal and Thiru A.P.C.V. Chokalingam, Secretary, V.O.C. College, Tuticorin, for providing the necessary facilities to do this work. Thanks are also due to our project staff Mr.S.Subramanian, Mr. S. Mariappan, Mr. T. Mali and Mr. S. Kadarkari for their assistance in the field and laboratory works.

References

Arora, C.L. and Bose, B.W., (1981) Demarcation of fresh and saline water zones using electrical methods-Abohar area, Ferozepur district, Punjab. J. Hydrology, V. 49, pp. 75-86.

Balasubramanian, A., Sharma, K.K. and Sastri, J.C.V., (1985) Geoelectrical and hydrogeochemical evaluation of coastal aquifers of Tambraparni basin, Tamil Nadu Geophy. Res. Bull, V. 223(4), pp. 204-09.

Stuyfzand, P.S., (1989) A new hydrochemical classification of water types. ISMA publ. 182, pp. 89-98.

Van Nostrand, R.C. and Look, K.L., (1986) Interpretation of resistivity data, U.S. Geol. Surv. prof. Paper No. 499.

40

Hydrogeological Studies in the Vaigai River Basin, Tamil Nadu

R. CHELLASAMY

AND

A. BALASUBRAMANIAN

Introduction

Vaigai River basin is located between latitude 9°15′ to 10°25′N and longitude 77°15′ to 79° covering an aerial extent of 8600 sq. km. in the Madurai and Ramanathapuram districts of Tamil Nadu. India (Fig. 40.1). The river Vaigai, originates at an altitude of 2200 m. above mean sea level in the western ghats, drains through the plains and confluences with the Bay of Bengal near Attangarai of Ramanathapuram district. It is sanctified by association with Lord Rama. The basin is bounded by western ghats in the west, Palni hills in the north, a stretch of mountain ranges comprising the Varushanad and Andipatti hills in the south and the Bay of Bengal in the east.

Physiography : Physiographically the basin could be classified into three divisions, *viz.*, 1. The highly elevated hills and ghats (existing in the western, south western and north western portions of the basin. 2. The gently sloping plains (covering the major portions) at the centre, and 3. the small stretch of the coastal belt (existing at the confluencing zone). The altitude in the hilly terrain ranges from 600 to 2400 m.

The general slope of the basin is towards SW-NE in the upper reaches and NW-SE in the lower reaches. The gradient of the coastal zone is very gentle. The landforms and land systems of the basin have been studied from LANDSAT (Black and White) images and false colour composites. The major land systems identified are : (*i*) hilly terrains : (*ii*) undulating terrains : (*iii*) rolling plains and (*iv*) fluvial plains.

Drainage : The geomorphometry of the basin has been studied on the basis of Horton (1945), Strahler (1953), Miller (1953), Schumm (1956) and Melton (1958). Morphometric analysis of the drainage network reveals that the basin encounters mostly dendritic to sub-dendritic drainages. The drainage density of the basin is found to be very low in most of the areas denoting the high potentiality of groundwater. The Vaigai river

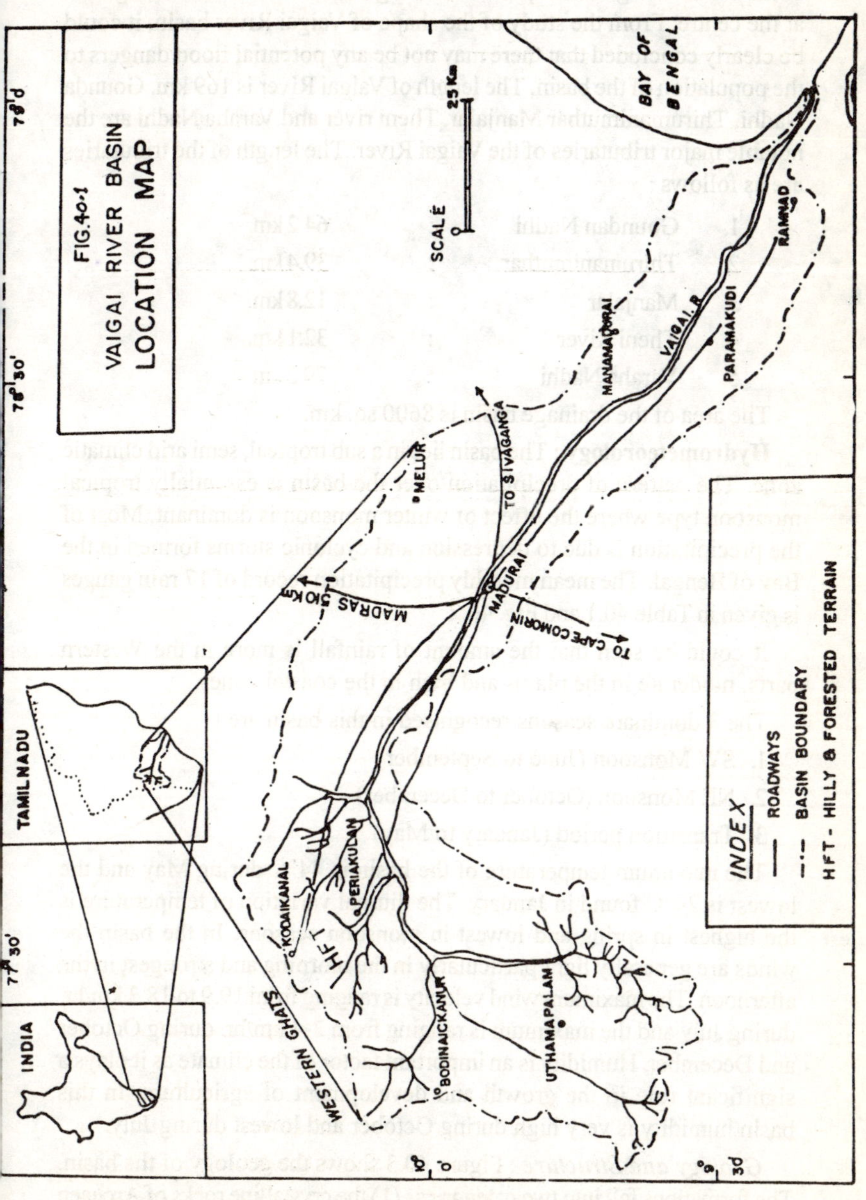

FIG. 4·1 VAIGAI RIVER BASIN LOCATION MAP

basin has an elongated shape with a change in the direction of elongation at the centre. From the study of the shape of Vaigai River basin, it could be clearly concluded that there may not be any potential flood dangers to the population of the basin. The length of Vaigai River is 169 km. Gounda Nadhi, Thirumanimuthar Manjalar, Theni river and Varaha Nadhi are the notable major tributaries of the Vaigai River. The length of the tributaries are as follows :

1.	Goundan Nadhi	:	64.2 km.
2.	Thirumanimuthar	:	39.4 km.
3.	Manjalar	:	12.8 km.
4.	Theni River	:	32.1 km.
5.	Varaha Nadhi	:	79.2 km.

The area of the drainage basin is 8600 sq. km.

Hydrometeorology : The basin lies in a sub tropical, semi arid climatic zone. The pattern of precipitation over the basin is essentially tropical monsoon type where the effect of winter monsoon is dominant. Most of the precipitation is due to depression and cyclonic storms formed in the Bay of Bengal. The mean monthly precipitation record of 17 rain gauges is given in Table 40.1 and Fig. 40.2.

It could be seen that the amount of rainfall is more in the Western parts, moderate in the plains and high in the coastal zones.

The 3 dominant seasons recognised in this basin are :

1. SW Monsoon (June to September)
2. NE Monsoon (October to December)
3. Transition period (January to May)

The maximum temperature of the basin is 44°C during May and the lowest is 24°C found in January. The diurnal variation of temperature is the highest in spring and lowest in monsoon seasons. In the basin the winds are generally light particularly in the morning and strongest in the afternoon. The maximum wind velocity is ranging from 19.9 to 18.3 km/hr. during July and the maximum is ranging from 2-4 km/hr. during October and December. Humidity is an important factor of the climate as it plays a significant role in the growth and development of agriculture. In this basin humidity is very high during October and lowest during July.

Geology and Structure : Figure 40.3 shows the geology of the basin. The formations fall into two categories : (1) the crystalline rocks of Archaen age in Madurai district and (2) the Recent to Sub-Recent sediments in Ramanathapuram district.

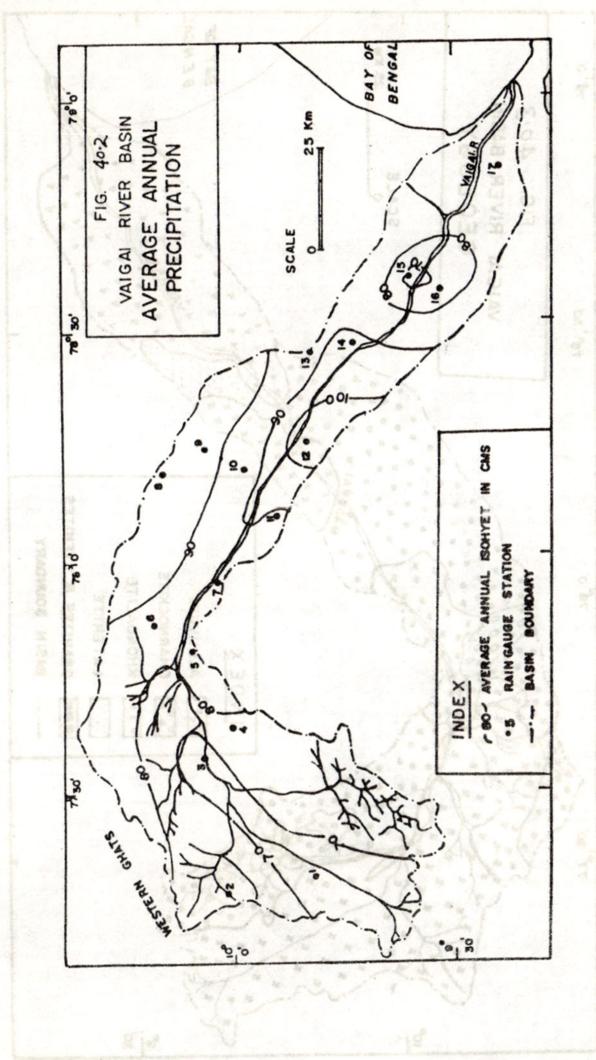
FIG. 40·2
VAIGAI RIVER BASIN
AVERAGE ANNUAL
PRECIPITATION
SCALE
0
25 Km
BAY OF
BENGAL
VAIGAI R.
WESTERN
GHATS
INDEX
80 AVERAGE ANNUAL ISOHYET IN CMS
•5 RAINGAUGE STATION
BASIN BOUNDARY

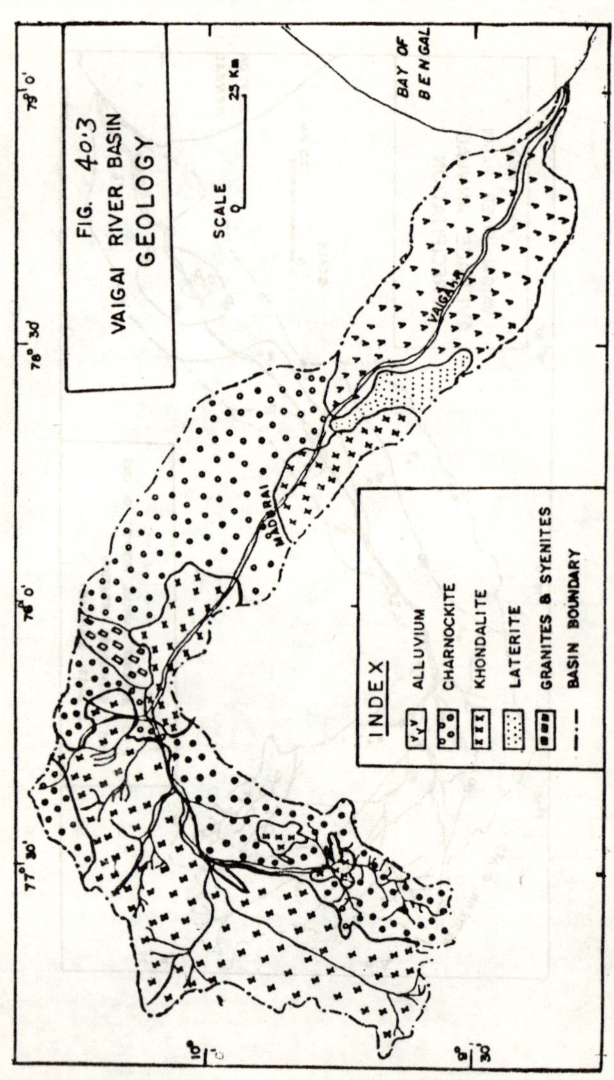

FIG. 40·3 VAIGAI RIVER BASIN GEOLOGY

Table 40.1 : Normal Rainfall in mm

Location name	*J*	*F*	*M*	*A*	*M*	*J*	*J*	*A*	*S*	*O*	*N*	*D*	*Total*	*Alt. mts.*
1. Thamapalayam	1	9	20	58	58	22	58	46	67	145	93	77	734	360.00
2. Bodinayakanur	23	6	34	84	81	16	40	36	83	191	111	71	776	345.00
3. Vaigaidam	12	10	20	52	60	23	41	25	51	184	129	56	663	186.00
4. Andipatti	1	1	28	29	53	27	29	60	171	210	114	51	774	225.00
5. Peraiyur	24	21	25	86	65	21	26	62	95	204	160	60	849	165.00
6. Nilakottai	10	26	30	68	87	60	60	64	179	175	82	57	898	265.00
7. Cholavandan	22	15	16	66	66	39	25	82	91	159	181	48	810	200.00
8. Nattam	32	12	16	70	71	52	54	123	130	215	149	66	990	256.00
9. Pulipatti	19	14	10	56	61	37	58	199	86	200	127	50	917	136.00
10. Chatrapatti	39	7	13	61	74	26	26	44	64	195	176	82	807	175.00
11. Thallakulam	24	18	22	69	68	35	40	91	89	198	152	60	866	154.00
12. Thiruppvanam	41	10	26	64	60	63	93	140	134	173	170	85	1232	35.00
13. Sivaganga	33	15	14	68	56	36	53	110	91	172	148	89	885	39.00
14. Manamadurai	37	16	26	62	47	46	58	111	99	166	153	94	915	90.00
15. Ilayankudi	7	6	10	47	10	39	30	78	48	157	139	94	665	9.00
16. Paramakudi	43	18	24	57	7	21	37	65	67	150	166	84	770	66.00
17. Ramanad	51	27	26	62	29	7	20	38	36	196	252	134	878	104.00

The Archaean complex comprises: (*i*) Charnockite and (*ii*) Khondalite rocks. The Charnockite group includes acid charnockites and related migmatites with bands of basic granulites and magnetite quartzites. These form the country rocks in Andipatti, Palani and Sirumalai hills. The Khondalite group consists of crystalline limestone, calcgneiss, clacgranulite, garent sillimanite-gneiss, hornblende biotite.

The formations existing in Ramanathapuram district are mainly of sedimentary rocks of upper Gondwana age consisting of basal boulder bed and conglomerates, micaceos sandstone, alternating shales and grits unconformably overlying the Archaean rocks. Hard calcareous tuffs and Kankar occur as a thin layer of about 35-50 cm over the weathered gneisses in many parts of the area.

Joints and fractures are found in the rock types like quartzites, gneisses, leptinites and granites. Charnockites show multiple sets of fractures and fissures which act as the pathways for groundwater movement.

Material and Methods

In order to assess the performance of wells in this terrain, aquifer characteristics with reference to charnockites, khondalites and sedimentaries have been evaluated using the pumping test data of 47 large diameter open wells. Zones of high transmissivity have been noticed in the lower reaches and river courses. The average time required for full recovery of dug wells in this basin has been determined. Potable and probable zones of further development have also been identified.

Many computer methods have been proposed and utilized during the last two decades. Reliable estimates of aquifer constants have been attempted through digital automated kit of the data and sensitivity analysis proposed by Mc Elwee (1980).

Results and Discussion

Hydrogeology

Based on the analysis of longterm water table data, specific yield and precipitation patterns, Loganathan and Kandasamy (1980) have determined the percentage of rainfall recharge for Tertiary (17%) weathered crystalline (10-12%) and for fissured and jointed rocks (8-10%). The average annual water level fluctuation is found to be more (>4m) in the northwestern portions of the basin and less at the downstream end (2-4 m).

Groundwater occurs under water table conditions in the crystalline terrain and most of the wells dry up during summer. The aquifer parameters of this basin have been evaluated using pumping and recovery test data of 47 dug wells (Table 40.2). Such studies help in delineating the potential zones for further groundwater development.

The Transmissivity (T) of aquifers is depicted in Fig. 40.4. Zones of high T have been noticed in the lower reaches of the basin. Presence of only a few pockets of high T horizons in the upstream side denotes that no further development could be done over these zones. However, areas adjacent to the river courses can be developed as they show a nominal range of T. The optimum yield (Karanjac, 1975) of the wells have been assessed (Fig. 40.5). It shows the nature of the aquifers both in the sedimentaries and in the crystalline formations.

Figure 40.6 shows the average time required for full recovery (tfr) of dug wells in this basin. The zones which need less than 12 hours and more than 48 hours of time for full recovery (Rajagopalan *et al.*, 1983) after complete abstraction, have been highlighted in order to adopt

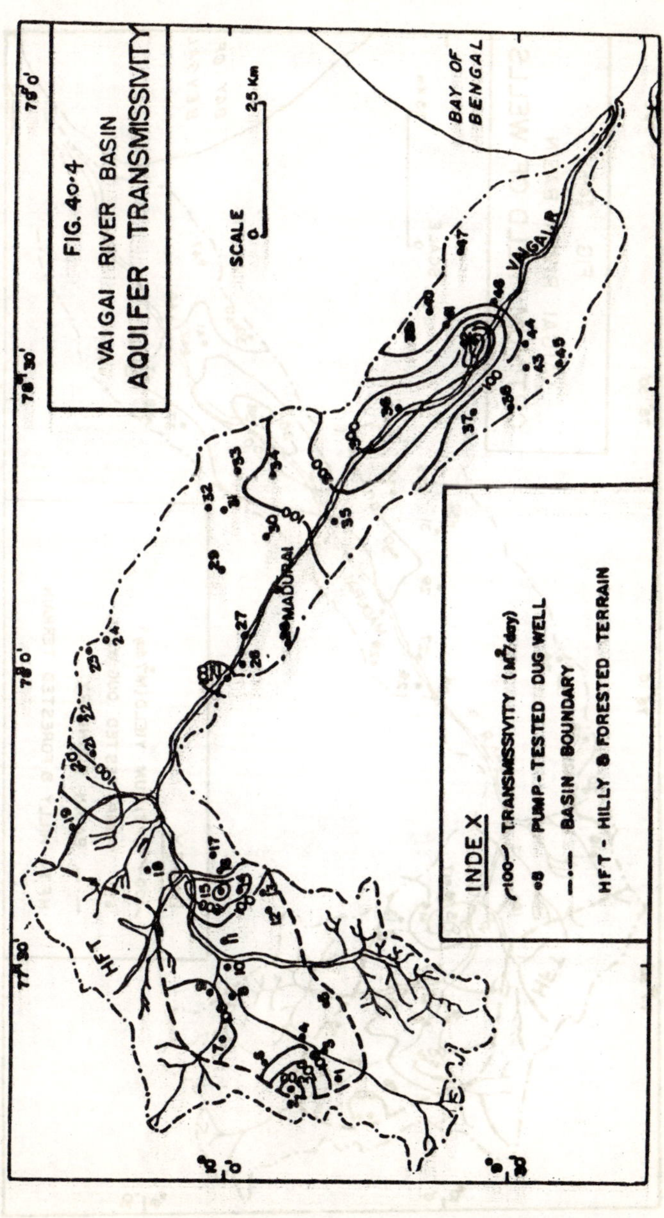
FIG. 40·4
VAIGAI RIVER BASIN
AQUIFER TRANSMISSIVITY
SCALE
0
25 Km
BAY OF BENGAL
VAIGAI R.
MADURAI
HFT
INDEX
100 TRANSMISSIVITY (M²/day)
•8 PUMP-TESTED DUG WELL
BASIN BOUNDARY
HFT - HILLY & FORESTED TERRAIN
78°0'
78°30'
77°30'

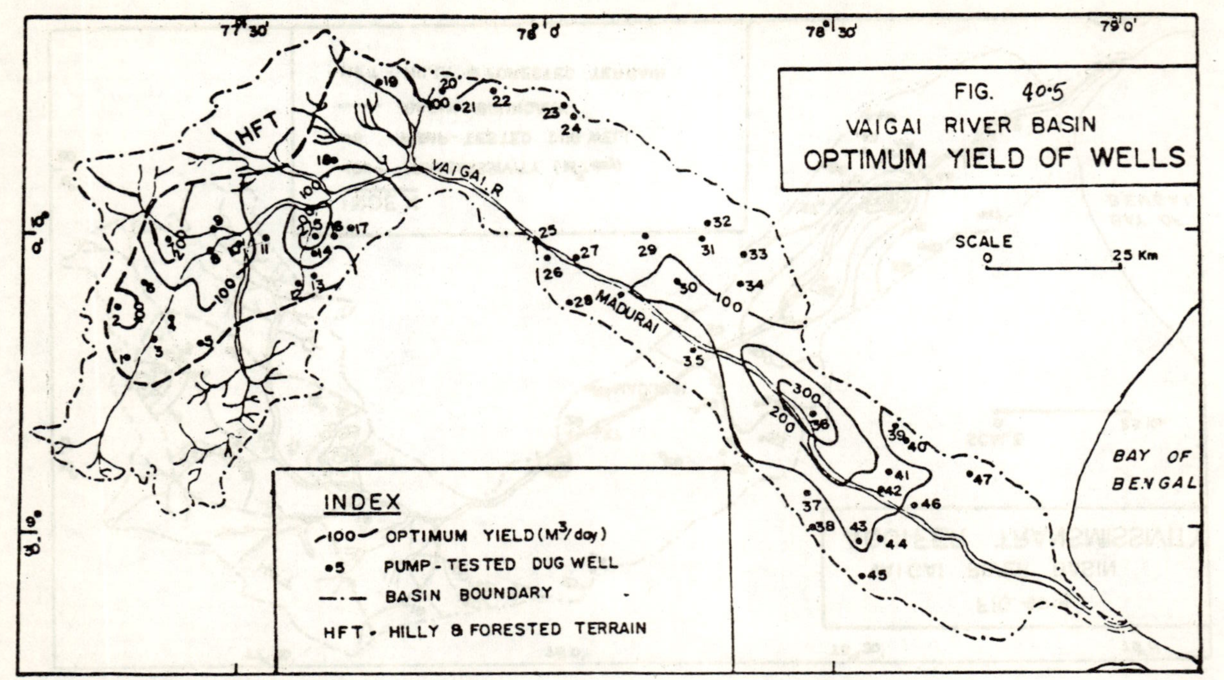
FIG. 40·5
VAIGAI RIVER BASIN
OPTIMUM YIELD OF WELLS
SCALE
0
25 Km
77°30'
78°0'
78°30'
79°0'
HFT
VAIGAI.R
MADURAI
BAY OF
BENGAL
INDEX
100 OPTIMUM YIELD (M³/day)
•5 PUMP-TESTED DUG WELL
BASIN BOUNDARY
HFT - HILLY & FORESTED TERRAIN

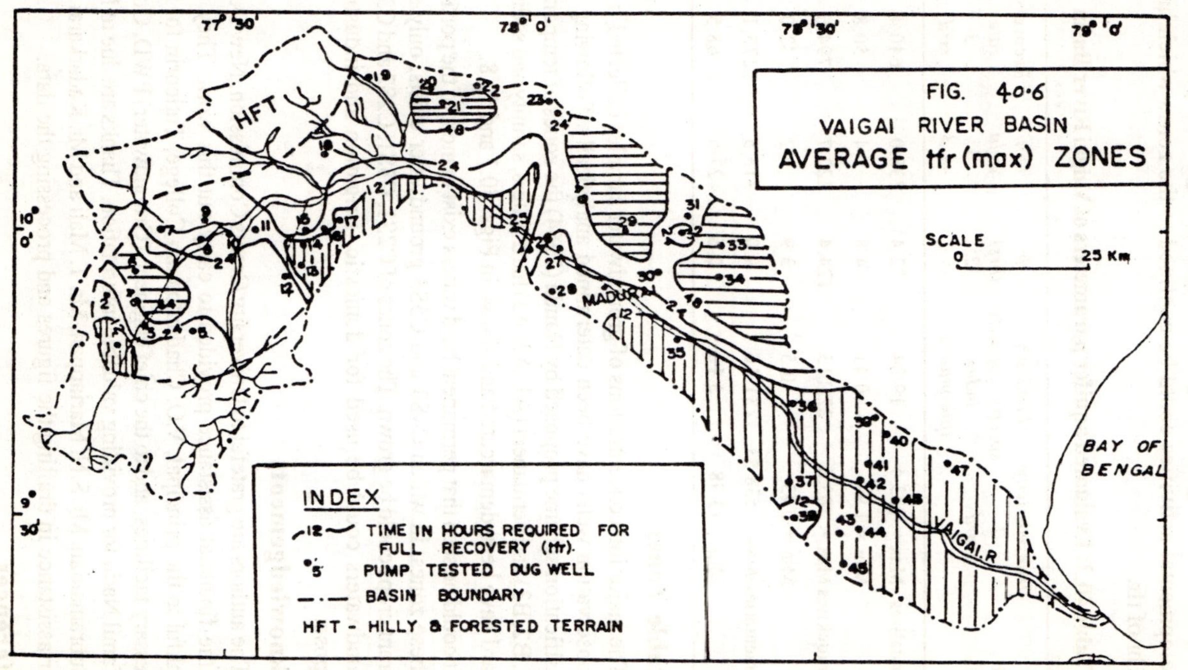
FIG. 40·6
VAIGAI RIVER BASIN
AVERAGE tfr (max) ZONES
SCALE
0 25 Km
BAY OF BENGAL
VAIGAI R.
MADURAI
HFT
INDEX
12 TIME IN HOURS REQUIRED FOR FULL RECOVERY (tfr)
5 PUMP TESTED DUG WELL
BASIN BOUNDARY
HFT HILLY & FORESTED TERRAIN
78° 0'
78° 30'
77° 30'
10° 0'
9° 30'

management strategies in agriculture and irrigation systems. All developmental activities could be confined to the zones which have <12 hr of tfr.

Table 40.2: Evaluated aquifer parameters of Vaigai River Basin

Rock type	*Transmissivity 2 m/d*	*Limaye's specific capacity Index ipm/mdd/m*	*tfr (hrs) 2*	*Opt. yield 3 m/d*	*Recovery rate 3 m/d*
Khondalites	Max = 333.4	19.04	72.4	238.0	640.4
	Min = 1.99	0.41	4.8	2.4	50.8
Charnockites	Max = 303.6	22.113	124.4	222.7	794.6
	Min = 0.43	0.5	3.9	0.2	42.0
Sedimentaries	Max = 538.53	231.16	9.0	313.5	1273.1
	Min = 13.38	5.16	0.6	27.5	68.5

Potable Zones

The major ion concentrations of groundwater samples collected from 68 observation wells have been determined analytically. Following the classification scheme proposed by Handa (1964) the zones of temporary (B1, B2, B3) and permanent (A1, A2, A3) hard waters, salinity and sodium hazard have been demarcated and shown in Figs. 40.7 and 40.8.

It could be seen that permanent hard waters exist in most of the portion. In these zones, in which C4S3 and C5S3 groundwater exists, only salt tolerant crops could be grown. The zones of C2S1, C3S1, C2S2 and C3S2 groundwaters could be used for domestic, irrigation and industrial purpose.

Acknowledgements

The authors are grateful to University Grants Commission, New Delhi for the financial assistance provided to carry out this work. They are thankful to the principal, V.O. Chadambaram College, Tuticorin for the necessary facilities, and to the chief engineer, (Groundwater) PWD, Govt. of Tamil Nadu for providing valuable information. Thanks are due to Mr. S. Subramanian, Mr. S.N. Mariappan, Mr. T. Mali and Mr. Kadarkarai for their assistance in drafting the figures and processing the data.

References

Handa, B.K., (1964) Modified classification procedure for rating irrigation water, Soil, Sci., V. 98, pp. 264-69.

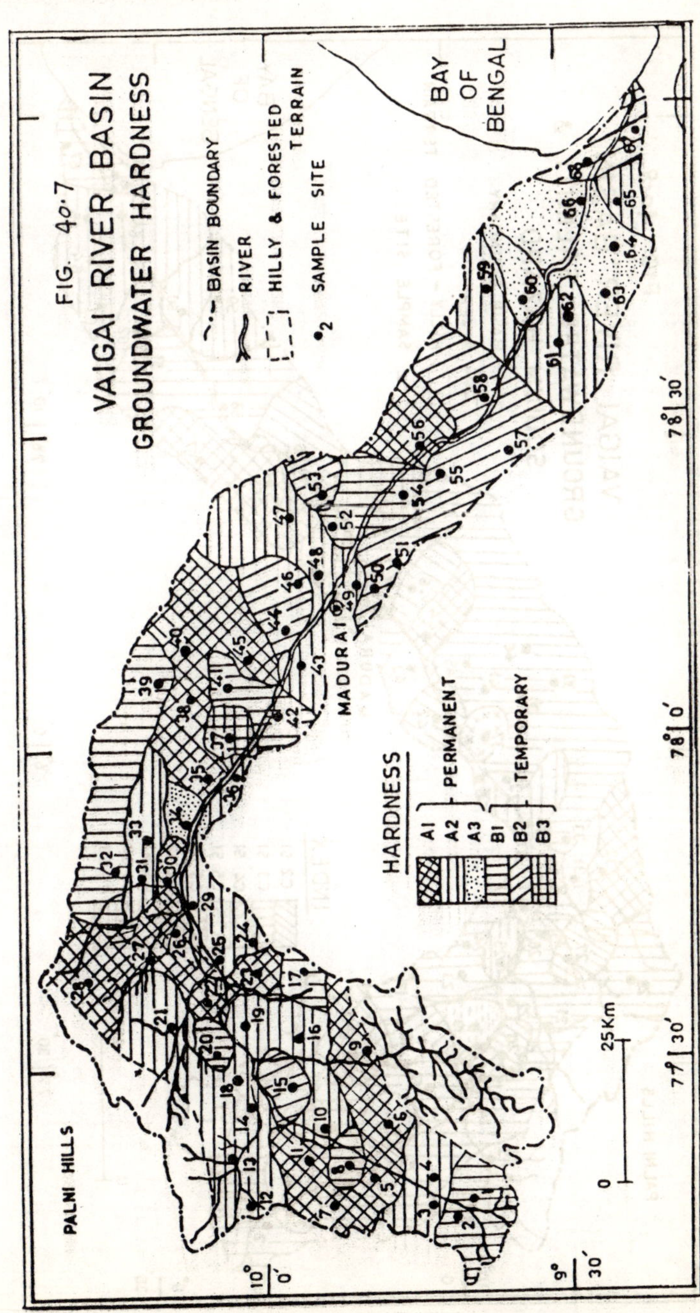
FIG. 40·7
VAIGAI RIVER BASIN
GROUNDWATER HARDNESS
BASIN BOUNDARY
RIVER
HILLY & FORESTED TERRAIN
SAMPLE SITE
BAY OF BENGAL
PALNI HILLS
MADURAI
HARDNESS
A1
A2
A3
PERMANENT
B1
B2
B3
TEMPORARY
0
25Km
10° 0′
9° 30′
77° 30′
78° 0′
78° 30′

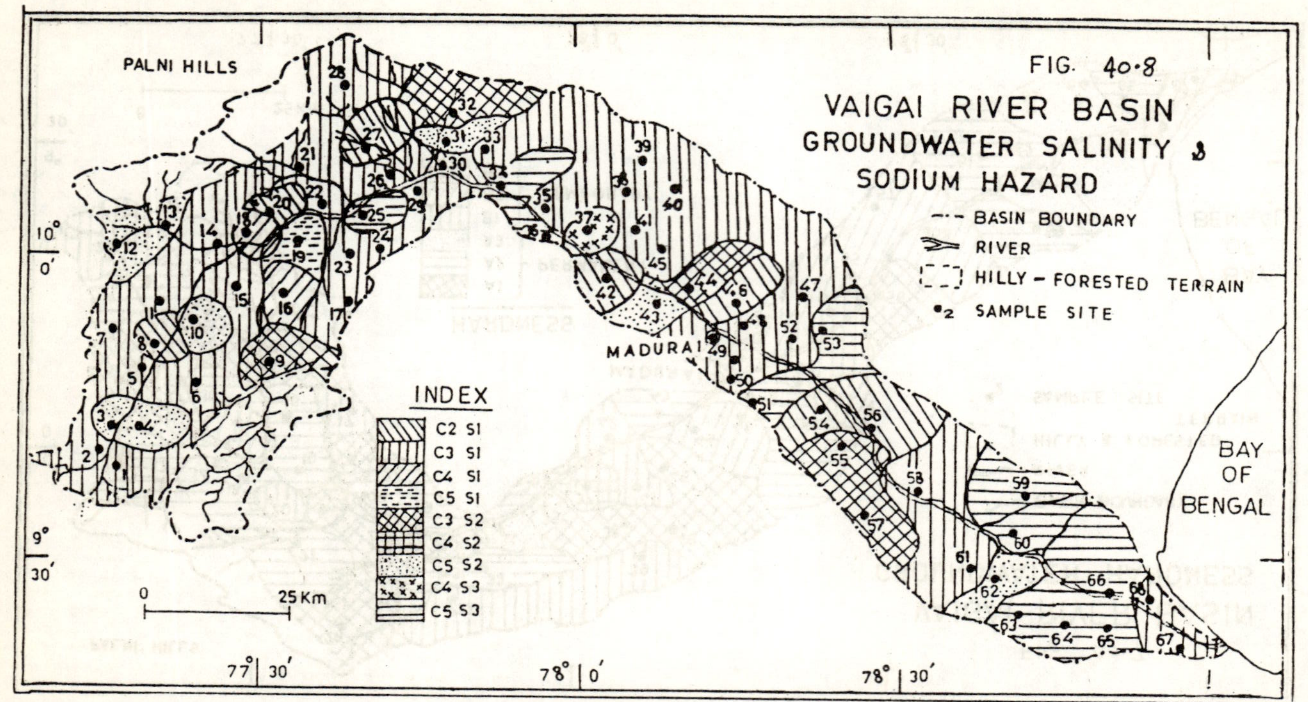
FIG. 40·8
VAIGAI RIVER BASIN
GROUNDWATER SALINITY &
SODIUM HAZARD
BASIN BOUNDARY
RIVER
HILLY – FORESTED TERRAIN
SAMPLE SITE
BAY OF BENGAL
PALNI HILLS
MADURAI
INDEX
C2 S1
C3 S1
C4 S1
C5 S1
C3 S2
C4 S2
C5 S2
C4 S3
C5 S3
0
25 Km
10° 0'
9° 30'
77° 30'
78° 0'
78° 30'

Horton, R.E., (1945) Erosional development of streams and their drainage basins, Hydrophysical approach to qualitative morphology, Geol. Soc. Amer. Bull. 56, 275-370.

Karanjac, J., (1975) Brief note on testing the possibility of defining optimum yield of dugwells of larger diameter, Groundwater, V. 13, pp. 4.

Loganathan, A. and Kandasamy, V., (1980) Correlation between rainfall and water level rise in selected control wells in Tamil Nadu. Proceeding of the Symposium on water resources of the southern states, Dept. of Geography, Madurai Kamaraj University, pp. 1-7.

Mc Elwee, C.D., (1980) The thesis equation, evaluation, sensitivity to storage and transmissivity and automated fit to pumptest data, Kansas Geol. Surv. Groundwater Ser-3, 39 p.

Melton, M.A., (1958) Correlation structure of morphometric properties of drainage systems and their controlling agents, Geol. V. 66, pp. 442-60.

Miller, V.C., (1953) A quantitative geomorphic study of drainage basin characteristic in the clinch mountain area, Virginia and Tennessee, Tech. rep. 3, Project NR. 389-342, Office & Naval Research Geogr. Branch, 30 p.

Rajagopalan, S.P., Abraham S. and Prabha Sankar, P.N., (1983) Pumping tests and analysis of test data from open wells in the coastal tract of Trivandrum District, Rep. GW/R-92/83, Groundwater Division, CWDRM, Calicut.

Schumm. A.S., (1956) Evaluation of drainage systems and slopes in badlands at Perth Amboy, New Jersey, Bull. Geol. Soc. Am., V. 67, pp. 597-646.

Strahler, A.N., (1953) Revision of Hortons quantitative factors in erosional terrains. Paper read before Hydrology section of the American Geophysics Union.

41

Hydrologic Cycle and its Effects on the Flowing Wells in the Rewa Area, Madhya Pradesh, India

YAMUNA SINGH

AND

D.P. DUBEY

Introduction

The area of present study is underlain by the Rewa Sandstone (Rewa Group) and the Ganurgarh Shale and the Bhander Limestone (Bhander Group) of the Vindhyan Supergroup (Precambrain) in ascending order. The contacts of all these lithounits are gradational. The Rewa Sandstone forms the hill ranges both in the south (known as Kaimur range) and in the north (known as Sohagighat range) (Fig. 41.1).

For optimum utilisation of groundwater for drinking and irrigation purposes, the investigations carried for water-well drilling programme launched by the Public Health Engineering Department (PHED) and the Lift Irrigation Corporation (LIC), Government of Madhya Pradesh, unfolded many facts about hydrogeological conditions in the Rewa area. Values of specific capacity of the wells have been found to vary from one formation to another. In limestone they vary from 88 litre per minute per metre (lpmin/m) to 1125 lpmin/m, in shale from 151 lpmin/m to 377 lpmin/m, whereas in sandstone they range from 155 lpmin/m to 378 lpmin/m. In general number of failures have been found more in shale (50%), followed by sandstone (35%) and limestone (25%).

During the course of water-well drilling programme, a number of flowing wells have been encountered in shaly terrain (Ganurgarh Shale) of the Rewa area. The objective of this paper is to highlight occurrence of flowing wells in the Rewa area of M.P., and to evaluate effects of hydrologic cycle on flowing wells.

Hydrometeorology, Climate and Rainfall

The available data on various components of the hydrologic cycle for Rewa area have been collected from different agencies, evaluated and presented diagramatically (Figs. 41.2, 41.3, 41.4 and 41.5). From the available data it is noticed that the climate of the area is tropical with hot

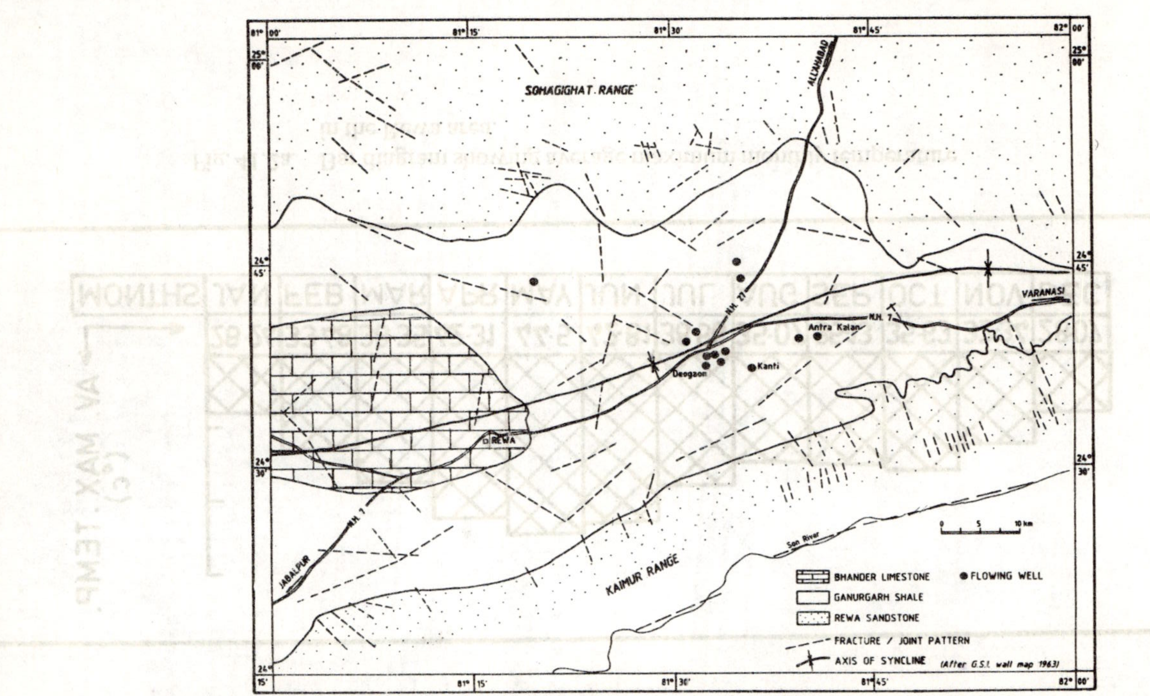

Fig. 41.1. Geological map of the Rewa area (based on photointerpretation and field checks) showing fracture/joint patterns, axis of syncline and locations of the flowing wells.

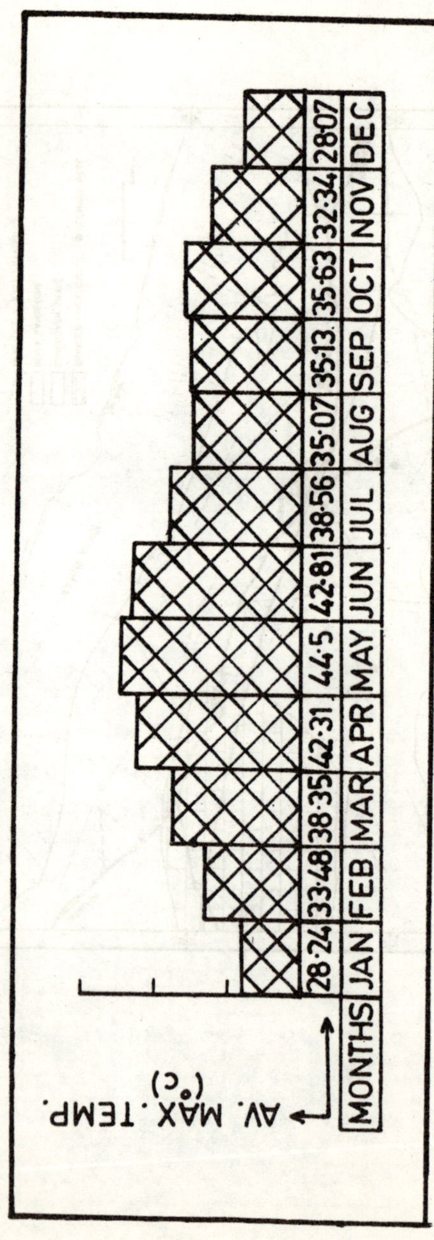

Fig. 41.2a. Bar diagram showing average maximum monthly temperature in the Rewa area.

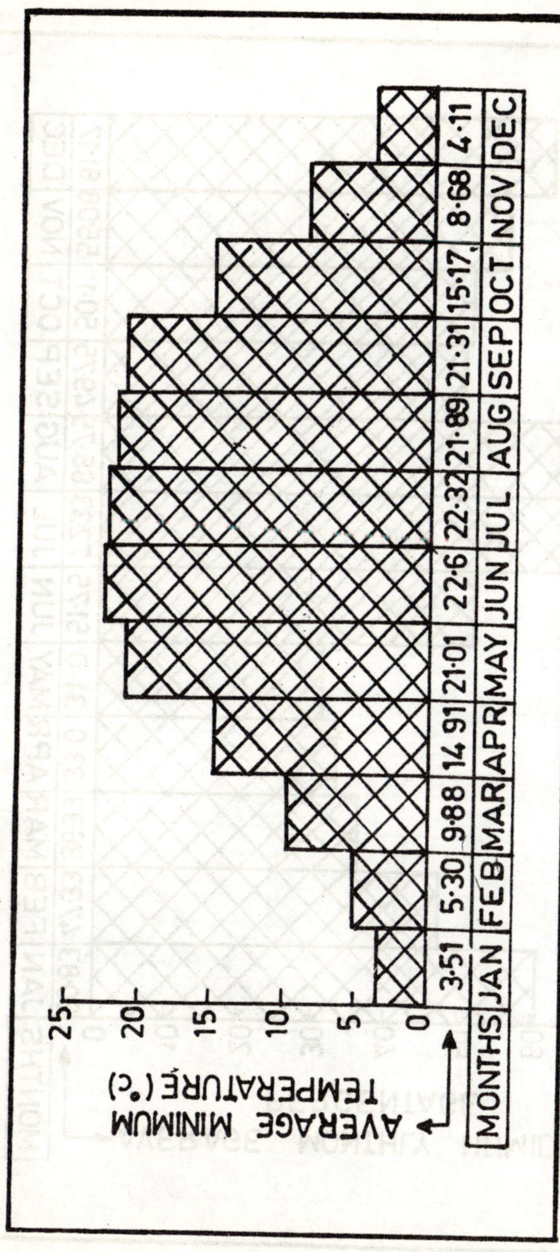

Fig. 41.2b. Bar diagram showing average minimum monthly temperature in the Rewa area.

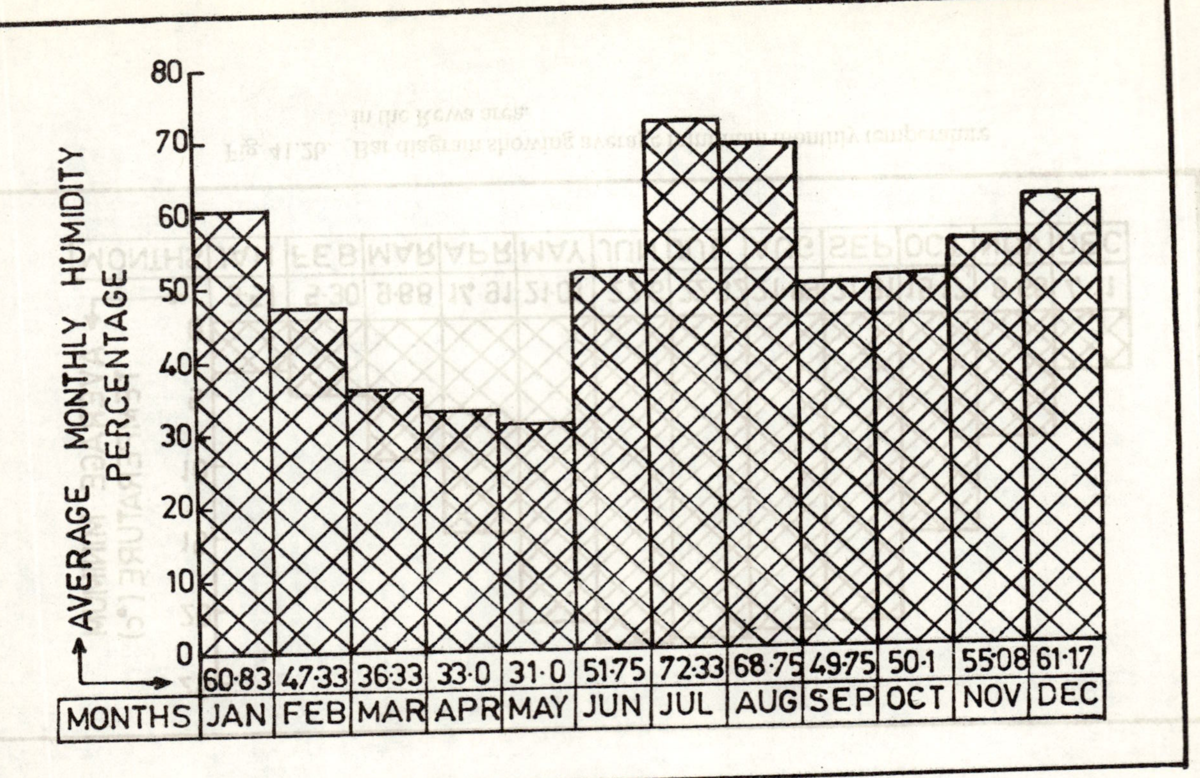

Fig. 41.3. Bar diagram showing average monthly relative humidity in the Rewa area.

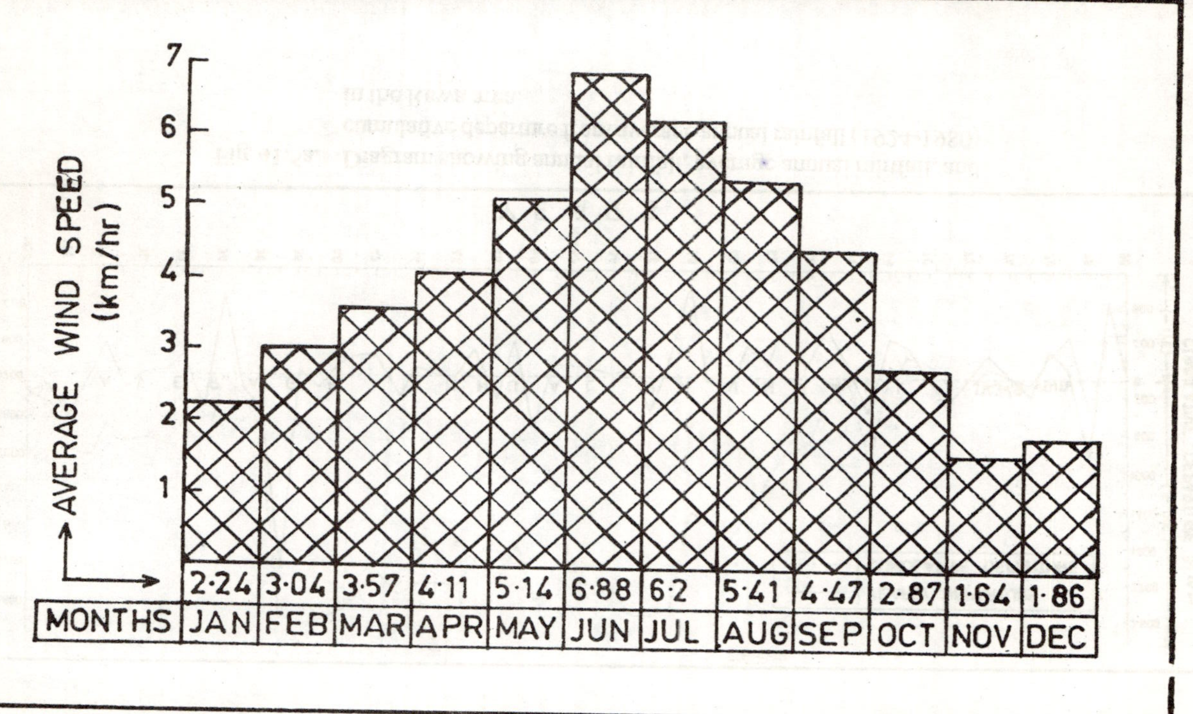

Fig. 41.4. Bar diagram showing average monthly wind speed in the Rewa area.

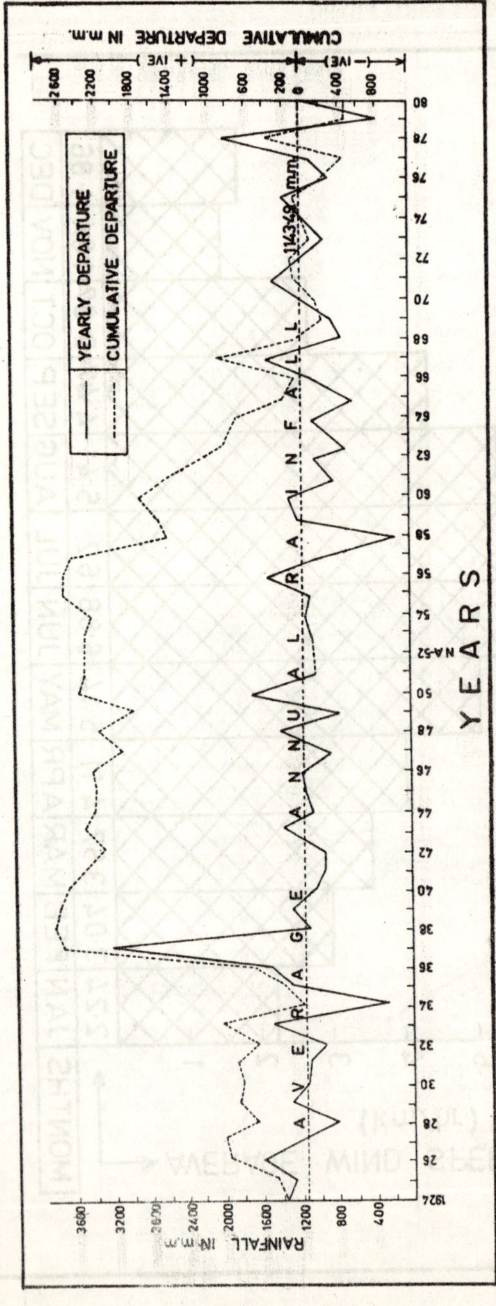

Fig. 41.5a. Diagram showing annual rainfall, average annual rainfall, and cumulative departure from average annual rainfall (1924-1980) in the Rewa area.

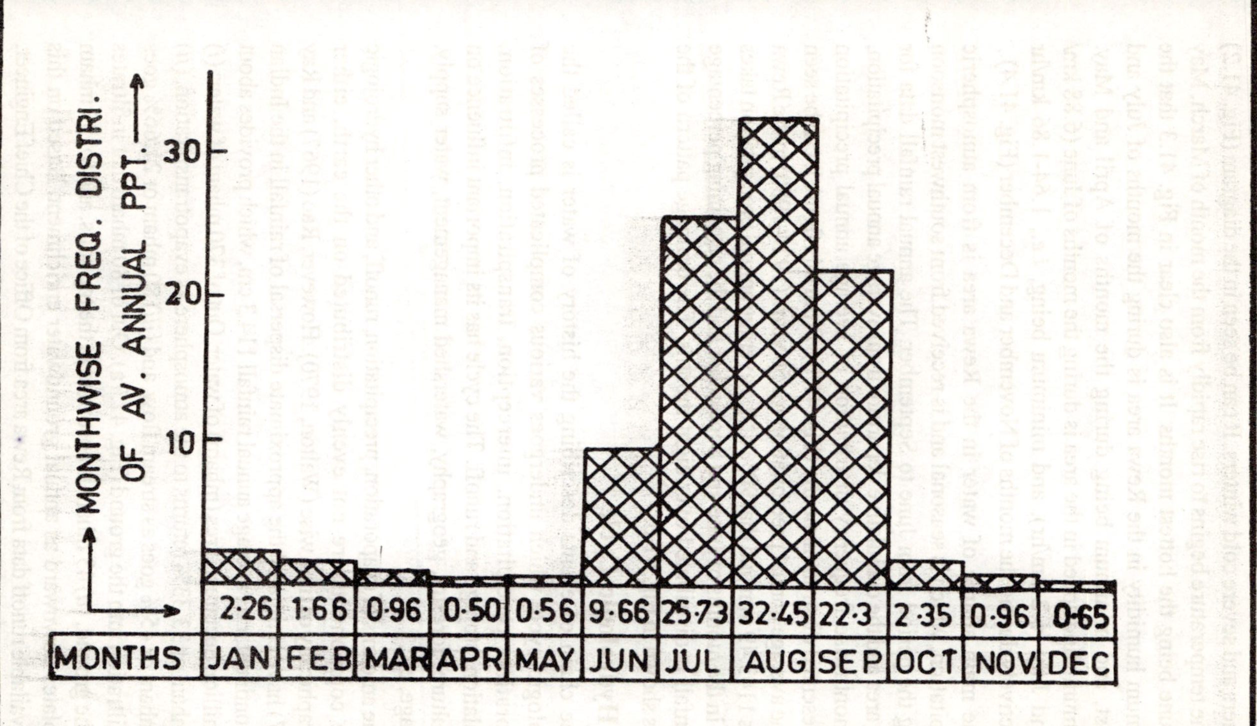

Fig. 41.5b. Bar Diagram showing monthwise frequency distribution of the average annual rainfall in the Rewa area.

summers and severe cold winters. It can be seen in the diagram (Fig. 41.2) that the temperature begins to rise rapidly from the month of March, May and June being the hottest months. It is also clear in Fig. 41.3 that the maximum humidity in the Rewa area is during the months of July and August, and minimum being during the months of April and May. Maximum wind speed in the area is during the months of June (6.88 km/hr) and July (6.2 km/hr), and minimum being, *i.e.*, 1.64-1.86 km/hr respectively during the months of November and December (Fig. 41.4).

The main source of water in the Rewa area is from atmospheric precipitation, which is seasonal and is received from southwest monsoon during the period from June to September. The annual rainfall data for Rewa area and the cumulative departure from average annual precipitation, and monthwise frequency distribution of average annual precipitation are presented diagrammatically (Fig. 41.5). From this figure it can be seen that the average annual precipitation for 56 years (1924-1980) in the Rewa area is 1143.49 mm and that 90% of the average annual precipitation takes place in the months of June and September and the remaining percentage of rainfall is distributed over eight months. The drainage pattern of the area is shown in Fig. 41.6.

The Hydrologic Cycle

The chain of events describing the history of water is called the hydrologic cycle which undergoes various complicated processes of evaporation, precipitation, interception, transpiration, infiltration, percolation, storage and runoff. The cycle has its important influence on agriculture, forestry, geography, watershed management, water supply, drainage, etc.

The amounts of evaporation, precipitation, runoff, and other hydrologic cycle components are not evenly distributed on the earth, either geographically or time wise (Walton, 1970). However, Rao (1967) and Ray (1967) have estimated the approximate dispersal of rainfall in the Indian sub-continent, *i.e.*, average annual rainfall 114.3 cm, which provides about 370 million hectare metres (mham) of water. Out of 370 mham of water: (*i*) 123 mham or 33.33% returns to the atmosphere as evapotranspiration, (*ii*) 167 mham or 45% goes as stream flow, and (*iii*) 80 mham or 21.66% goes as infiltration into the ground (Fig. 41.7a). Out of 80 mham that infiltrates into the ground, (*a*) 43 mham is absorbed in the top soils, and (*b*) 37 mham percolates downward as annual groundwater enrichment. Based on this and available runoff data from Rewa area from Office of the Chief Engineer, Ganga Basin, Rewa, approximate percentages of runoff, infiltration and

Fig. 41.6. Topographic/drainage map of the Rewa area, Madhya Pradesh.

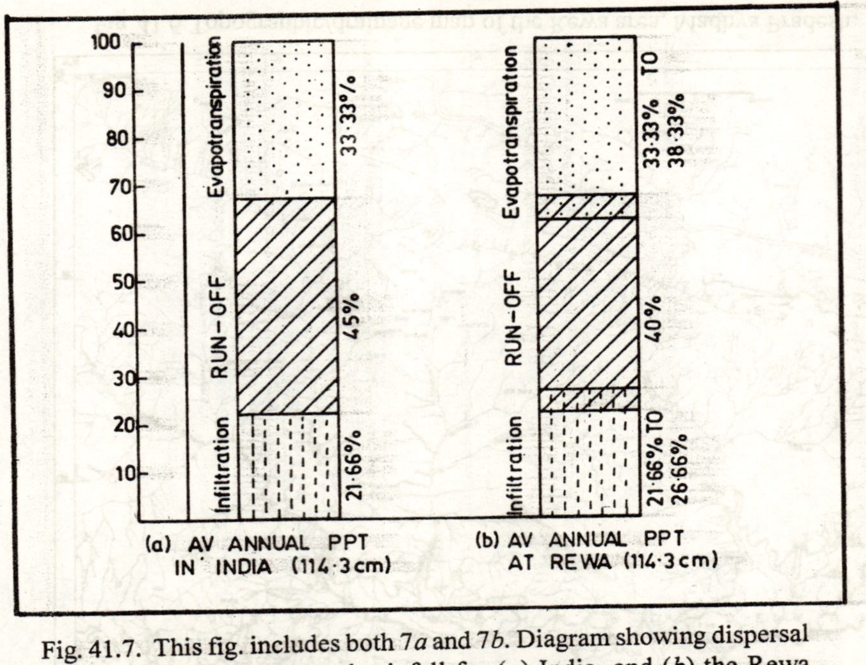

Fig. 41.7. This fig. includes both 7*a* and 7*b*. Diagram showing dispersal of average annual rainfall for (*a*) India, and (*b*) the Rewa area.

evapotranspiration for the Rewa area have been estimated as below and presented diagrammatically (Fig. 41.7*b*).

(*i*)	Average annual rainfall for 56 years (1924 to 1980)	:	114.35 cm.
(*ii*)	Evapotranspiration	:	38.1 to 43.8 cm (33.33 to 38.33%)
(*iii*)	Runoff	:	45.73 cm
(*iv*)	Infiltration	:	24.76 to 30.47 cm (21.66 to 26.66%)

The Flowing Wells

During the course of water-well drilling programme by PHED and LIC, a number of flowing wells have been encountered in shaly terrain of the Rewa area. Some of them are located around Delhi (24°44'00"N; 81°19'35"E), Paharkha (24°46'10"N; 81°35'10"E), Deogaon 24°37'35"N; 81°32'15"E), Kanti (24°37'45"N; 81°35'40"E) and Antra Kalan (24°43'30"N; 81°41'50"E). In all these areas when the production wells were first drilled in the bed rock, they flowed with hydrostatic heads ranging from 0.5 to 3.0m above surface. In some of the cases, however, flow was not accompanied by significant heads above the land surface. At Antra Kalan hydrostatic head was found maximum, *i.e.*, about 3.0m above the land surface, which gradually went on reducing with the passage of time. At Kanti (Fig. 41.8) when a well was drilled first it flowed with a head of about 1.5m above the land surface, which also slowly went on declining with the passage of time. It may also be noted here that in Kanti area drilling of 8-10 production wells within the radius of about 1km. caused lowering of the water level in wells significantly.

Effects of Hydrologic Cycle on the Flowing Wells

According to Aller, *et al.*, (1987), depth of water level varies from over 30m near the mountains to zero in the discharge areas. In the Rewa area depth of water level from the surface in the wells varies with respect to topography and lithography. It can be seen from Fig. 41.9 that during premonsoon period, depth to water level increases from alluvium (average 4.40 m) to limestone (average 5.50 m) through sandstone (average 8.36 m) and shale (average 8.75 m).

From well inventory data it is observed that there is significant rise in water level during rainy season. It is interesting to note that the water level fluctuations in different formations are not uniform (Fig. 41.9). Net rise in water level being minimum in alluvium (average 3.50m), followed

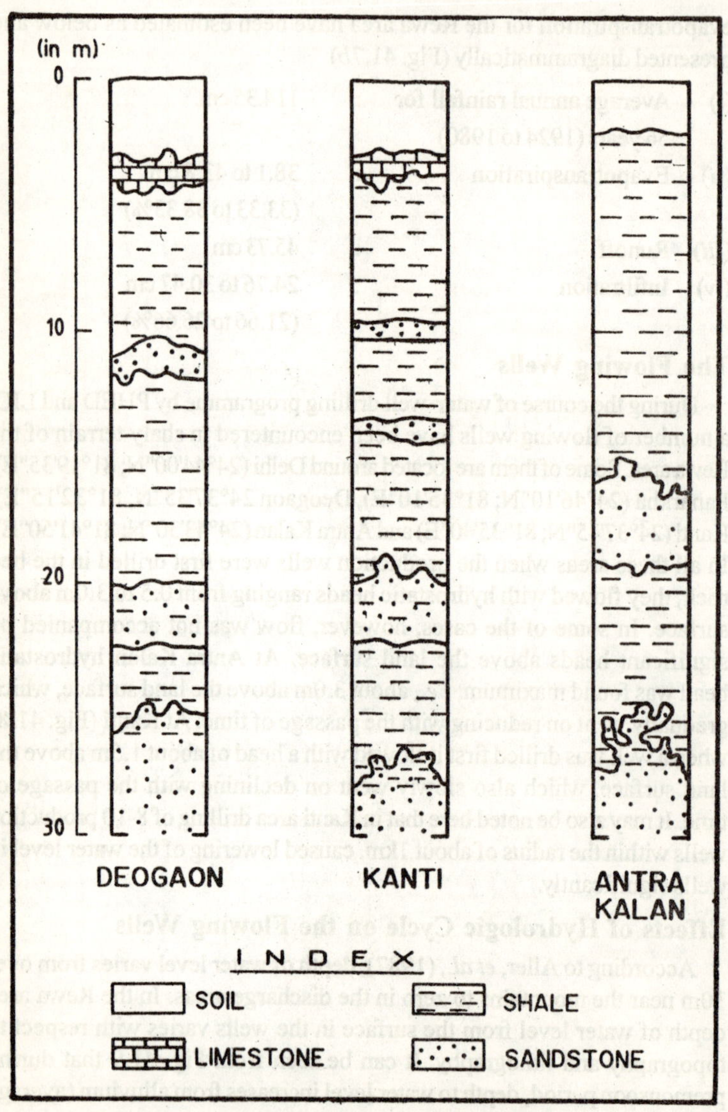

Fig. 41.8. Litholog data at the selected locations of flowing wells in the Rewa area, Madhya Pradesh.

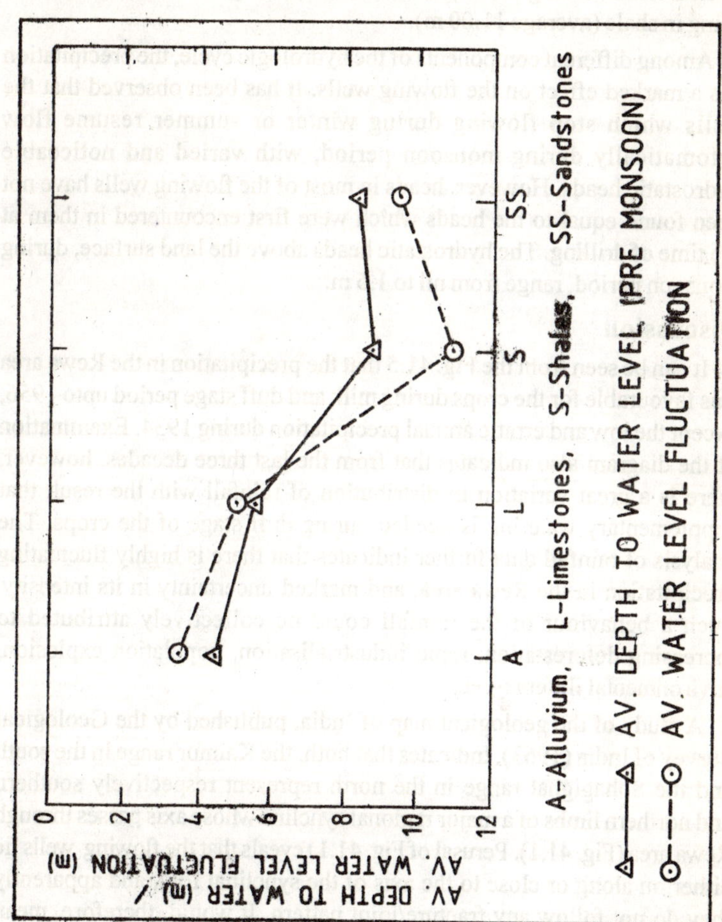

Fig. 41.9. **Diagram showing average depth of water levels during premonsoon period, and average water level fluctuations during rainy season in alluvium, limestones, shales, and sandstones of the Rewa area.**

by limestone (average 5.10 m), sandstone (average 9.40) and maximum being in shale (average 11.00 m).

Among different components of the hydrologic cycle, the precipitation has a marked effect on the flowing wells. It has been observed that the wells which stop flowing during winter or summer resume flow automatically during monsoon period, with varied and noticeable hydrostatic heads. However, heads in most of the flowing wells have not been found equal to the heads which were first encountered in them at the time of drilling. The hydrostatic heads above the land surface, during monsoon period, range from nil to 1.5 m.

Discussion

It can be seen from the Fig. 41.5 that the precipitation in the Rewa area was favourable for the crops during milk and duff stage period upto 1956, except the low and erratic annual precipitation during 1934. Examination of the diagram also indicates that from the last three decades, however, there is a great variation in distribution of rainfall with the result that supplementary watering is needed during duff stage of the crops. The analysis of rainfall data further indicates that there is highly fluctuating precipitation in the Rewa area, and marked uncertainty in its intensity. Such a behaviour of the rainfall could be collectively attributed to increasing deforestation, rapid industrialisation, population explosion, environmental illiteracy, etc.

A study of the geological map of India, published by the Geological Survey of India (1963), indicates that both, the Kaimur range in the south and the Sohagighat range in the north represent respectively southern and northern limbs of a major regional syncline whose axis passes through Rewa area (Fig. 41.1). Perusal of Fig. 41.1 reveals that the flowing wells lie either on along or close to the axis of the synclinal fold, and apparently they do not follow any fracture/joint pattern. It would, therefore, mean that it is mostly regional synclinal fold that controls the localisation of the flowing wells in the Rewa area.

Examination of the core samples indicated the presence of sandwiched sandstone karstified limestone lenses and beds of pinching and swelling nature in shale sequence. Thus it is quite probable that dense shales act as aquitards over the permeable sandwiched lenses and simulate confined conditions. Impermeable nature of the shale and pinching out of the sandstone and limestone lenses and body seem to be the prime reasons for more failures of the wells and not encountering flowing wells throughout the shaly terrain.

Review of Fig. 41.9 indicates wide variations in water level fluctuations, during pre- and post-monsoon periods, in different formations. Minimum water level fluctuations in alluvium and maximum in shales are attributable respectively to their high and low specific yields. Phenomenon of lowering of water levels with outgoing of the monsoon, and their rise during monsoon period causing flow in wells are due to limited areal extent of the aquifers and groundwater recharge mostly during monsoon period.

Taking the annual calculated average rainfall 114.35 cm, and infiltration as 26% of the rainfall (30 cm) for the Rewa area, the approximate amount of recharge into the ground from infiltration will be 3,00,000/m^3 per sq. km. area. There may be recharge of groundwater from base flow from outside the area and also from the rivers of the area during the flood periods. The amount of infiltration in limestone terrain is likely to be more because they are karstified (Singh, 1985; 1992). This inference is supported by observations of Burdon and Dounas (1961), who calculated 45.2% infiltration into the limestone aquifer, and 3.6% surface runoff for the Parnassoc-Ghaina area of Greece.

Conclusions

(*i*) In the Rewa area of Madhya Pradesh, the average annual rainfall for 56 years (1924-1980) is 114.35cm, and annual estimates of evapotranspiration, runoff and infiltration are 38.1 to 43.8cm, 45.73cm, and 24.76 to 30.47cm respectively.

(*ii*) Regional synclinal fold appears to be responsible for localisation of the flowing wells in shaly terrain of the Rewa area. The permeable sandstone and karstified limestone lenses and beds of pinching and swelling nature act as aquifer, whereas shales act as aquitards over the permeable lenses and simulate confined conditions.

(*iii*) There is increasing order of water level fluctuations, due to precipitation, from alluvium to limestone-sandstone to shale, in response to decreasing specific yields of these formations. There is marked effect of the atmospheric precipitation on the flowing wells; the wells, which stop flowing during winter or summer, resume flow automatically during monsoon period, in response to rise in water level, due to recharge of the aquifer from rainfall. The aquifers have got limited areal extent and are recharged mostly during rainy season.

(*iv*) The approximate amount of recharge, in the Rewa area, into the ground from infiltration annually per square kilometre will be 3,00,000m^3.

Acknowledgements

The authors are indebted to authorities of the PHED and LIC, Rewa, M.P., for their keen interest in the work and providing bore hole cores and yield data of the wells. Meteorological data of the Rewa area were provided by Indian Meteorological Department, Government of India, Nagpur, and runoff data for the same area by Office of the Chief Engineer, Ganga Basin, Rewa, for which the authors are thankful to them. Shri S.N. Chaurasia and Shri N.K. Muthukrishnan are thanked for drawing figures. An anonymous reviewer is thanked for useful suggestions.

References

Aller, L., Jay, H.L.,Petty, R. and Bennett, T., (1987) Drastic: A standardized system to evaluate groundwater pollution potential using hydrogeologic setting. Jour. Geol. Soc. India, V. 29, pp. 23-37.

Burdon, D.J. and Dounas, A., (1961) Hydrochemsitry of the Parnassos-Ghiana aquifers, and problems of sea-water contamination in Greece. Paper on Groundwater Resources, UNESCO.

Rao, K.L., (1967) Irrigation foundation for sound economy. Conqueror, V. 2, No. 23 (Weekly Magazine, Madras).

Ray, A., (1967) Water in ground.Jour. Andhra Pradesh Academy of Science, V. 1, pp. 29-35.

Singh, Y., (1985) Hydrogeology of the Carstic area around Rewa, M.P., India In : Karst Water Resources (Proceesings of the Ankara Antalya Symposium, July 1985). IAHS Publ. No. 161, pp. 407-16.

Singh, Y., (1992) Study on the cave passages in Recambrain Bhander lime stone of the Vindhyan super-group, India, Z. Geomonph, N.F. Suppl.-Bd. 85, Berlin Stuttgart pp. 115-26.

Walton, W.C., (1970) Groundwater Resources evaluation. McGraw Hill Kogakusha, Ltd., 664 p.

42

Management of Poor Quality Groundwaters for Crop Production in India

S.K. Gupta

Introduction

Agriculture is the major activity in India and irrigation plays a preeminent role in increasing the production and in reducing the unevenness in year-to-year production related to variations in weather patterns. Although, both surface and groundwater have been used throughout the ages, rivers have been the obvious and traditional source of water. Major efforts in irrigation development have been made to harness this resource, following the severe drought of 1966-67, when the legendary myth of dependence on surface irrigation systems was shattered and their utter vulnerability to such situations was exposed; concerted efforts were made to develop the groundwater resource. Presently, important role of the groundwater in the growth of agriculture is well recognized.

Management of groundwater revolves around adjusting pumping rates total withdrawls due to constraints imposed by aquifer properties and long term average of renewable recharge. Additionally, in arid and semi-arid regions, salinity of groundwater often imposes severe constraints in its development. In the latter cases, development of groundwater has been restricted because suitability of saline water for crop production has been misassessed and very conservative standards have conventionally been used in recommending groundwater development. Moreover, conventional criteria and procedures of producing irrigated crops with saline water have been considered. Climate, soil type, varietal differences in crop tolerance,crop substitution, irrigation techniques and conjunctive use techniques have often been ignored in making such recommendations, which might allow the use of more saline water for crop production without affecting the yield or the soil health.

The present paper has been prepared as a state-of-the art paper with the objective to colate evidences of the successful use of saline waters for irrigation in India and use such evidences for assessing crop tolerance to irrigations with saline waters. It is proposed to discuss strategies for greater and efffective use of saline water in crop production. These strategies include conjunctive use of saline and fresh waters, applying saline water only at the tolerant stages of growth and improved irrigation practices. In the last category is included a pitcher irrigation technique which is an effective alternative to the drip irrigation for conserving water and for allowing use of saline water for crop production.

Extent of the Problem

Although no precise estimate of the groundwater potential of the country are available, as a first approximation, potential of renewable groundwater is estimated at 40 M h a-m. How much of this resource is saline is yet to be estimated. Gupta (1979) made an extensive review of the water quality in various states of India and concluded that groundwater quality problem is quite serious in arid and semi-arid regions. Based on this survey and on the basis of additional information, Manchanda (1990) indicated a high percentage of saline groundwaters in Rajasthan and Haryana. In seven states for which data could be complied, 50 per cent of the groundwater have been found to be marginal or saline (Table 42.1).

Table 42.1 : Quality distribution (%) of underground waters in seven states of India

State	*Good*	*Marginal*	*Poor*
Punjab	59	22	19
Haryana	35	10	55
Rajasthan	16	16	68
Uttar Pradesh	37	20	43
Madhya Pradesh	75	10	15
Gujarat	70	20	10
Karnataka	65	10	25
Average	50	15	35

Crop Tolerance and Water Quality

Crop tolerance to salt stress has been assessed on the basis of soil salinity (Maas and Hoffman, 1977; Maas, 1986). Soil exchangeable sodium percentage (Pearson and Bernstein, 1958; Gupta and Sharma 1990), specific ions such as Boron (Maas, 1986) and Sodium (Maas, 1986; Rhoades, 1983). Maas and Hoffman (1977) visualised and proposed a model for describing crop tolerance which states that crops do not suffer yield reduction unless a threshold value is reached beyond which yield declines linearly with increasing soil salinity. Prichard *et al.*, (1983) suggested that the estimating suitability of a crop to irrigation with saline water; information on its tolerance to soil salinity should be known and a relation between soil salinity and irrigation water quality need to be developed. Such relations depend upon a a number of factors and therefore, are location specific and at the same location vary from an year to another (Prichard *et al.*, 1983; Gupta 1985). Similar limitations restrict the direct use of groundwater quality in the model, nevertheless, in this paper irrigation water quality has been treated as a variable directly for assessing crop response to irrigations with saline waters.

Description of Model

Piecewise linear model of Maas and Hoffman (1977) is written as:

$$Y = \begin{cases} 100 & 0 < ECiw/ECt \\ 100 - S(ECiw - ECt) & ECt < ECiw/ECo \\ 0 & ECiw > ECo \end{cases} \quad (1)$$

Here Y is the relative yield with respect to conditions where nonsaline water was used for irrigations; ECiw is the quality of the irrigation water in terms of its electrical conductivity; ECt is the electrical conductivity at or below which there is no yield reduction compared to non-saline water and S is the slope. The slope represents the yield reduction per unit increase in the electrical conductivity of the irrigation water. The parameter ECo is the electrical conductivity at or beyond which yield will be practically zero. Irrigation water quality at which pre-determined decrease in yield will occur can be assessed by the relation.

$$ECiw = ECt + \frac{\text{Desired yield reduction (\%)}}{S(\%)} \quad (2)$$

Crop Response to Irrigation with Saline Water

Crop response expressed in terms of the threshold and slope based on electrical conductivity of the irrigation water are given in Table 42.2. Electrical conductivity of the irrigation water at which yield is expected to be 50 per cent of the yield in comparison to non-saline conditions are also reported in Table 42.2. Values of EC 50 should in general indicate the relative potential of the crops with saline waters at a particular site and also at different sites provided one could take into account the variations in local conditions (Table 42.3).

Table 42.2 : Threshold irrigation water quality, slope and EC50 of selected crops

Crop	Soil texture	Location	Piecewise linear model		
			ECt	*Slope*	*EC50*
			(%)		
Wheat	Sandy loam	Agra	7.0	4.50	18.1
	Loamy sand	Jodhpur	3.5	8.90	9.1
	Sandy loam	Karnal	7.9	8.70	13.6
	Dune sand	Karnal	11.1	2.90	28.3
	Sandy loam	Hisar	3.4	5.50	12.5
	Sandy loam	Sampla	1.7	1.75	30.0
	Sandy loam	Kanpur	1.0	4.60	11.9
	Black clay	Dharwad	1.0	4.60	11.9
	Black clay	Indore	4.4	4.70	15.0
Barley	Sandy loam	Agra	8.3	2.20	31.0
	Sandy loam	Jobner	10.8	3.95	23.5
Rice	Sandy loam	Karnal	1.3	7.57	7.9
	Black Clay (S)	Baptla	1.0	7.79	7.4
	(W)		1.0	12.52	5.0
Maize	Black clay	Indore	1.0	7.17	8.0
	Black clay	Dharwad	1.0	2.84	18.6
Sorghum	Sandy loam (F)	Agra	3.2	2.96	20.1
	(G)	Agra	6.2	3.82	19.3
	Black clay	Dharwad	1.0	5.80	9.6
Pearl millet	Sandy loam (576.0)	Agra	4.6	2.03	29.3
	(408.0)		2.2	3.92	14.9
	(358.8)		4.1	6.75	11.6
Mustard	Sandy loam	Agra	6.6	7.35	13.4
	Sandy loam	Jobner	2.0	1.96	27.5
Sunflower	Sandy loam	Agra	4.9	7.82	11.3
	Sand (W)	Baptla	1.0	3.76	14.3
Safflower	Sandy loam	Agra	3.8	3.31	18.9
	Black clay	Dharwad	1.0	2.41	21.8
Groundnut	Sand	Baptla	1.0	11.54	5.3
Pigeon pea	Sandy loam	Agra	0.6	4.50	13.0
Gluster bean	Sandy loam	Jobner	2.0	7.77	8.4

Table 42.2 Contd.

Crop	*Soil texture*	*Location*	*Piecewise linear model*		
			ECt	*Slope*	*EC50*
Onion	Sandy loam	Agra	0.6	12.75	4.6
	Sand	Baptla	4.6	11.87	8.8
Lady's finger	Sandy loam	Agra	0.6	12.75	4.6
Fenugreek	Sandy loam	Jobner	2.0	5.59	10.9
Berseem	Sandy loam (F)	Agra	2.2	13.90	5.8
Itallian millet	Black clay	Dharwad	8.8	11.30	13.3
Finger millet	Black clay (Sarda)	Baptla	1.0	5.25	10.6
Potato	Sandy loam	Agra	1.7	6.26	9.7
Brinjal	Sandy loam (P)	Karnal	9.8	10.60	14.5
	Sand	Baptla	1.0	7.10	8.0

S = summer; W = winter; F = fodder; G = grain, P = Pitcher irrigation Numbers in parenthesis indicate average rainfall

Table 42.3 : Annual rainfall and distribution pattern at different stations

Location	*Average annual rainfall (mm)*	*Rainfall pattern**
Agra	660	E4(C3D1)E4
Baptla	900	E4(C3D1)C2E2
Dharwad	820	–
Hisar	410	E4(C2D1E1)E4
Indore	730	E4(B2C2)E4
Jobner	600	E4(C2D1E1)E4
Jodhpur	370	E4(C2D1E1)E4
Karnal	720	E4(C3D1)E4
Pali	410	E4(C2DIE1)E4

*As per National Commission on Agriculture Report, 1976.

In general crop response curves vary on account of factors such as rainfall, crop variety (Table 42.4), irrigation technique and level of management. At a particular site, crops could tolerate relatively more saline waters in light than a heavy textured soil. It is due to the fact that soil salinity, as a result of application of saline water, is more in a heavy than in a light textured soil. With increasing rainfall ECt increase. For example, it increased from 5.1 ds/m to 12.9 ds/m in case of wheat as the rainfall increased from 17.0 mm to 117.0 mm. Slope, however, increased initially and then decreased probably because of the fact that medium amounts of rainfall brought down salts right in the root zone affecting the crop more adversely.

Approaches for the use of Saline Water

There could be several approaches to develop appropriate system for the use of saline water for crop production. Important of those put into practice are:

(1) Direct use of saline water

(2) Conjunctive use of saline and fresh waters:

 (*a*) Blending saline and fresh water,

 (*b*) Cyclic use,

 (*c*) Avoiding sensitive stages of crop for irrigation with saline water,

(3) Improved irrigation technology,

1. Direct Use of Saline Water

Agricultural crop species differ significantly in their tolerance to excessive concentration of salts in the root zone. Thus, the performance of crops will also vary as a consequence of application of saline water. In fact the close relation between the two is that more the concentration of salts in the irrigation water more will be soil salinity, provided other factors remain unaltered. Crop response curves for wheat and barley-the two winter season crops and rice-a summer season crop for a sandy loam soil are reported in Fig. 42.1. Evidence from the existing information indicates that pearl millet-another summer season crop and cotton are more tolerant than rice and could replace rice in saline water agriculture. Thus, there is a great scope for adjusting crops in relation to availability of water and its quality. From the results in Table 42.2, it could be inferred that when salinity of the irrigation water exceeds 10 ds/m, it is advantageous to shift to barley in medium textured soils of the semi-arid regions. On the other hand if the water quality is relatively less and water availability is restricted,

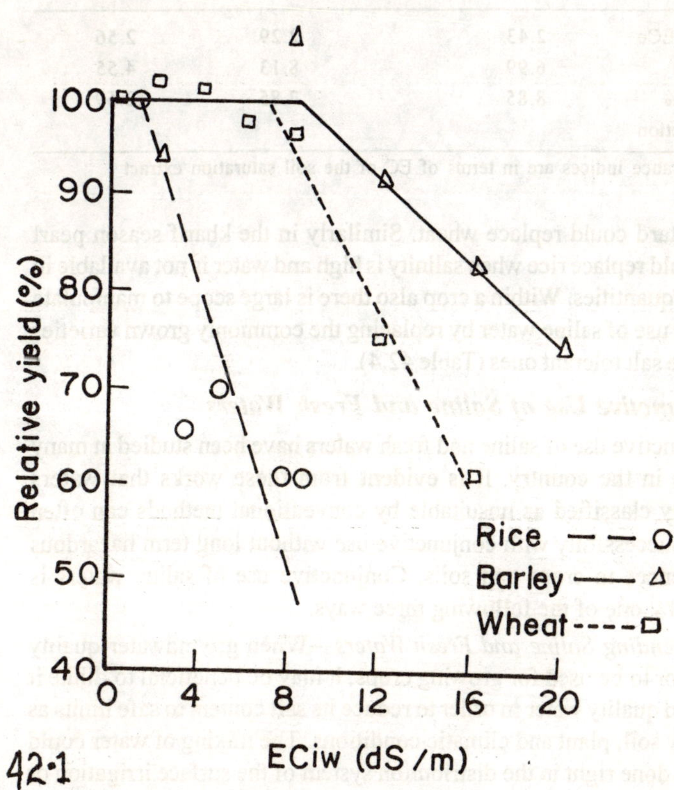

42·1

Crop response curves as related to salinity of irrigation water for 3 main crops of India

Table 42.4 : Effect of crop variety on salt stress in case of Wheat

*Salt tolerance indice**	*HD 4502*	*HD160*K65*	*K65*
Threshold ECe	2.43	2.29	2.56
Slope	6.99	8.13	4.55
ECe at 50% yield reduction	8.85	7.85	12.69

* Salt tolerance indices are in terms of EC of the soil saturation extract

then mustard could replace wheat. Similarly in the kharif season pearl millet could replace rice when salinity is high and water is not available in adequate quantities. Within a crop also there is large scope to manipulate the direct use of saline water by replacing the commonly grown varieties with more salt tolerant ones (Table 42.4).

2. *Conjunctive Use of Saline and Fresh Waters*

Conjunctive use of saline and fresh waters have been studied at many locations in the country. It is evident from these works that waters commonly classified as unsuitable by conventional methods can often be used successfully with conjunctive use without long term hazardous consequences to crops and soils. Conjunctive use of saline waters is possible by one of the following three ways.

(a) Blending Saline and Fresh Waters—When groundwater quality is too poor to be used for growing crops, it may be beneficial to dilute it with good quality water in order to reduce its salt content to safe limits as set out by soil, plant and climatic conditions. The mixing of water could either be done right in the distribution system of the surface irrigation or at the farm level. The work carried out in the state of Haryana (India) indicated that depending upon the quality and discharge of the fresh and saline waters, condition of the water course (Lined/Unlined), alignment of the water course and velocity of the falling water; distance of the mixing varied from 25 m to 60 m with an average of 42m (HSMITC, 1984). Mixing ratio should be such that resultant water quality should not exceed 0.74 s/m in case water at the down stream side is to be used for drinking purposes. In other cases, depending upon the cropping pattern, the resultant salinity could be increased to 1.0 to 2.0 d/m. Tyagi and Tanwar (1986) indicated a great potential of this technique in the state.

In spite of the great potential that this technique offers for the use of saline waters, many limitations have been recognized. For example, it may restrict either the choice of crops or germination of several crops may be affected. It may also cause social tension between the farmers at the head-end and at the tail-end. Additionally Rhoades (1983) reported that mixing of water may not add to the useable water resource for crop production, if highly saline waters are used in mixing. In the present study it puts such a contention into an inequality as follows :

$$Qm(1\text{-}LF_m) > Qf(1\text{-}LF_1)$$

As long as this inequality is satisfied, mixing will be useful. In this relation Q is the quantity of water and LF is the leaching fraction necessary to restrict salt built up beyond the desired level. The subscripts m and f denotes that mix water and fresh water respectively. The calculations show that to get a mix water of 1.0 ds/m from fresh water of 0.5 ds/m with a LFf of 0.055, blending with saline waters up to 8 ds/m will only be useful. Beyond this value, there will not be any resource addition.

(*b*) *Cyclic Use*—The premise of this strategy is based on two facts, i.e., (1) good quality water in arid and semi-arid regions is in short supply and (2) the salt tolerance of crops vary from a stage to another. According to this strategy, saline water is substituted for good water either when fresh water is not available and/or at times when the crops at a more tolerant growth stage. The timing and amount of water application will depend upon the quality of water, the cropping pattern, the climate and the type of the irrigation system. The crop yields with canal water, tube well water and with cyclic use indicate that one irrigation with canal water followed by one irrigation by the tube well is at par with canal water alone (Bajwa *et al.*, 1987) in rice-wheat rotation (Table 42.5). Cyclic use of rainfall with saline water has resulted in the tolerance indices to improve. It was reported that 117.0 mm of rainfall (Nearly equal to 2 irrigations) could permit more saline waters for production of wheat at the same yield level compared to when rainfall is only 17.0 mm.

Number means number of irrigations, for example, 2g-1s means two irrigation with good water followed by 1 irrigation by saline water

sw had an EC = 1.3 dS/m, SAR = 13.5 and RSC = 10 meq/l

(*c*) *Avoiding Sensitive Stages During Irrigations with Saline Waters*—Several modifications to cyclic use of saline waters have since been tried. The basis of these modifications is to reduce the requirement of the fresh water and apply it only at the most sensitive stage of growth. For example, researchers have shown that many crops are highly sensitive at the germination or initial establishment stage. Therefore, if fresh water is applied at this stage, more saline water can be applied at the later growth stage of the crop. It has been observed that in case of Indian mustard yield reductions were 12 and 54 per cent when saline waters of 7.9 ds/m and 12.3 ds/m respectively were used. The yield reduction declined to 5 and 22 per cent when pre-sowing irrigation was given by fresh water of 0.8 ds/m and saline waters were used thereafter (Minhas *et al.*, 1988).

Table 42.5 : Average grain yield (t/ha) of crops as affected by cyclic use of saline/sodic (sw) and good water (gw)

Treatment	*1st Year*		*2nd Year*		*3rd Year*	
	Rice	*Wheat*	*Rice*	*Wheat*	*Rice*	*Wheat*
gw	7.01	5.86	6.80	5.53	6.57	4.94
sw	4.91	3.82	3.73	3.49	3.63	1.95
2g-1s	6.89	5.77	6.77	5.05	6.35	5.01
1g-1s	6.51	6.13	6.47	4.85	5.97	5.05
1g-2s	6.43	5.69	5.60	4.53	5.14	4.34
CD at 5%	0.55	0.66	0.78	0.44	0.45	0.48

Fallowing of the land is yet another technique commonly followed in India. The winter crop is grown with saline water but land is left fallow during the summer seasons. The excess rainfall during the monsoon season leaches down the salt making the land fit for cultivation. It is estimated that nearly 80 per cent of the salts are leached below the root zone in areas where rainfall is around 400 mm. A dual rotation system is yet another possibility which has met with success (Rhoades, 1983). According to this system, a relatively tolerant crop is grown with saline water while in the following season a less sensitive crop is grown with fresh water. The accumulated salts are leached down the profile during this season and process could be continued. Cotton-wheat rotation has been successfully tried with this technology. There appears to be a need to test these technologies for more crops and at different locations.

3. Improved Irrigation Technology

Irrigation technique could play an important role in modifying crop response behaviour when saline water is used for irrigation. Low application depths at high frequency helps in overcoming the effect of osmotic stress. It is now well known that sprinkler or drip irrigation could permit relatively more saline waters for crop production than any surface irrigation methods. In the Indian context, an indigeneous method of irrigation, *i.e.*, 'Pitcher Irrigation Technique' has been developed and tried at Central Soil Salinity Research Institute, Karnal (Montal *et al.*, 1987). It has performed excellently well for growing vegetables with saline waters. It is an excellent tool to conserve water where water resources is limited. In a nutshell, this technique consists of burying the earthen baked pitchers in the soil and fill these pitchers with water at regular intervals. The moist zone crated around these pitchers is sufficient to grow four plants of any vegetable crop (Fig. 42.2). The yield per pitcher for some crops is reported in Table 42.6. The water salinity at which there was no yield reduction and at which 50 per cent yield reduction occurred are also reported in this table. The values at which 50 per cent yield reduction occurred are relatively much higher than commonly reported salinities in case of surface irrigation methods.

Table 42.6 : Crop performance with saline water in Pitcher Irrigation Technique

Crop	*Yield (Kg/Pitcher)*	*ECt (dS/m)*	*EC50 (dS/m)*
Tomato	5.8	15.0	*
Cauliflower	5.2	15.0	*
Brinjal	5.1	5.2	15.3
Ridgegourd	4.5	3.2	15.0
Cabbage	4.8	5.0	17.5

* Not evaluated

Conclusions

From what has been stated in this paper, it appears that there is a great scope for the utilization of saline groundwaters in arid and semi-arid regions. In many irrigation commands, where water-logging and soil salinity threatens agricultural production, such techniques could be usefully applied for the reuse of saline drainage waters generated through irrigation return flow. This should help in solving the problem of water-logging and soil salinity on the one hand and boost agricultural production on the other particularly in the arid and semi-arid regions of the country. It is however emphasized that the technologies described herein should be critically examined, combined if necessary and tested for location specific situations to make them relevant for that region.

Acknowledgement

Author is grateful to Dr. N.T. Singh, Director Central Soil Salinity Research Institute, Karnal for the encouragement and help extended during the period of this study.

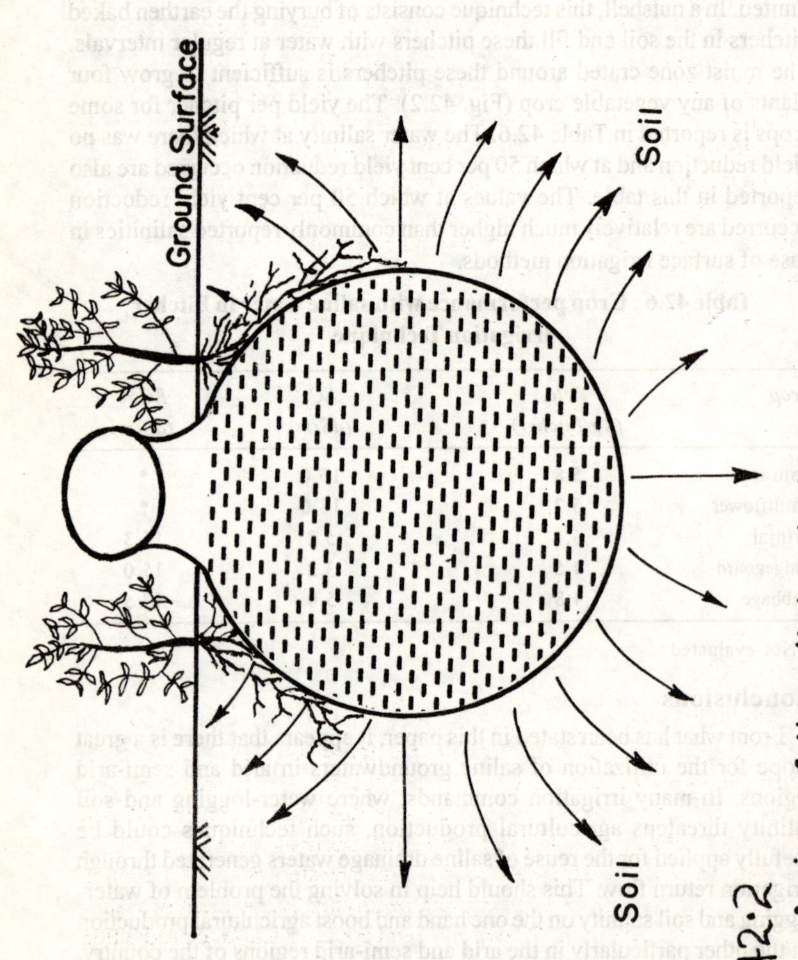

Fig.42.2 A view of the pitcher irrigation technique for conserving water and use of saline water

References

HSMITC, (1984) Appraisal document on studies for the use of saline waters in the command areas of irrigation projects, Haryana. Haryana State Minor Irrigation Tubewells Corporation (HSMITC), Chandigarh.

Bajwa, M.S., Sandhu, B.S. and Prihar, S.S., (1987) Soil and water management Problems of the South-West Punjab in relation to crop production. Presented at Water-logging and Soil Salinity Problems in South-West Punjab : Causes and Remedial Measure, Chandigarh.

Gupta, I.C., (1979) Use of saline water in agriculture in arid and semi-arid zones of India. Oxford and IBH Publishing Company, New Delhi.

Gupta, S.K., (1985) Dynamics of salts in saline water irrigated soils. J. Inst. of Eng. (India), V. 66, (ACI). pp. 17-22.

Gupta, S.K. and Sharma, S.K., (1990) Response of crops to high exchangeable sodium percentage. (Accepted, Irrigation Science).

Maas, E.V., (1986) Crop tolerance to saline soil and water. Proc. US-Pak. Biosaline Research Workshop, Karachi, Pakistan, pp. 205-12.

Maas, E.V. and Hoffman, G.J., (1977) Crop salt tolerance : Current assessment. J. Irrigation and Drainage Division, ASCE, V. 103, No. IR2. pp. 115.

Manchanda, H.R., (1990) Use of saline waters in agriculture. (Personal Communication).

Minhas, P.S., Sharma, D.R. and Khosla, B.K., (1988) Evaluation of irrigation Practices for crop utilizing saline waters. Annual Report, Central Soil Salinity Research Institute, Karnal. pp. 44-46.

Mondal, R.C., Gupta, S.K. and Dubey, S.K., (1987) Pitcher irrigation, Better Farming Series in Salt Affected Soils No. 11, Central Soil Salinity Research Institute, Karnal. pp. 15.

Pearson, G.A. and Bernstein, L., (1958) Influence of exchangeable sodium on yield and chemical composition of plants (2) wheat, barley, oats, rice, tall fescue and tall wheat grass. Soil Sci. V. 86, pp. 254-59.

Prichard, T.C., Meyer, J.L., Hoffman, G.J., Kegal, F.R. and Roberts, R., (1983) Relationship of irrigation water salinity and soil water salinity, California Agriculture, V. 37, No. 7-8, pp. 11-14.

Rhoades, J.D., (1983) Using saline water for irrigation. Proc. Int. Workshop on Salt-Affected Soils of Latin America, Maracay, Venuzuela.

Tyagi, N.K. and Tanwar, B.S., (1986) Planning for saline ground water resources. Proc. 53rd Annual R & D Session, Central Board of Irrigation and Power, Bhubaneshwar, P. No. 184, No. II, pp. 15-28.

Spatio-Temporal Analysis of Irrigation : A Case Study of Belgaum District, Karnataka

A.S. RAYAMANE

AND

S.S. NAREGAL

Introduction

*Belgaum district situated in the north-western part of the Karnataka State, falls within the northern maidan region. The district covers an area of 13,410 sq. Km. and consists of ten talukas. It has an average annual rainfall of 130 to 160 cm. and the temperature variation from 37.6°C to 41.5° (in summer) to 9.7°C to 12.8°C (in winter). The district has an area of 9,18,650 ha (66.96%) as net sown and 1,99,392 ha (20.61) as net irrigated (1984-85). Except Khanpur taluk and some parts of Ramdurg, Bailhongal and Sonndatti taluks most of the area of *Belgaum district is covered with black soil. The remaining part is covered by red and lateritic soil. The forest area was 1,92,776 ha (14.16%) during 1984-85. As per 1981 census, the district has 29,74,861 population, ranking second in the state of Karnataka.

In the preceding analysis an evaluation of irrigation in *Belgaum district has been made in detail in order to understand the role and impact of irrigation on development of agriculture, productivity and its efficiency.

Impact of Irrigation on Development of Agriculture

The study on irrigation will certainly contribute to a greater extent to understand the future course of probable change in land use of *Belgaum district and thereby need for planning of agriculture in the district. There are three important rivers (Krishna, Ghataprabha and Malaprabha), several streams, many tanks and wells in the district, most of them are utilised for irrigation purposes. But this utilisation is inadequate for the standing crops even in main cropping seasons, which has made the district to suffer from food grain shortages. As per the figures made available by the Bureau of Economics and Statistics. The *Belgaum district has an area of 1,99,392 ha of land under net irrigated during 1984-85 but it was 1,14,575 ha during 1974-75 which constitutes 21.34 present and 12.70 per cent of net area sown respectively . The net irrigated area has increased to 84,817 ha during a decade. The increase in net ir'rigated area over a decade is not uniform in all the taluks of the district. However, the increase in the net irrigated area in the district is a positive fact in the agricultural development.

The spatial distribution of net irrigated area during 1974-75 shows that, it was maximum in Raibag taluk, i.e., 30,855 ha which constituted 51.90 per cent of the taluks net sown area. It was followed by Gokak (27,727 ha and Athani taluk (19,514 ha) by constituting 24.90 and 11.50 percentage to their net sown area respectively. In remaining seven taluks irrigated area varies from 2,185 ha (in Soundatti taluk) to 7,569 ha (in Chikodi taluk). During 1984-85, the spatial distribution of irrigated area was maximum in Raibag taluk (35,994 ha) which constitutes 51.35 per cent of the of taluk's net sown area. In case of Gokak (37,367 ha), Soundatti (23,436 ha), Ramdurg (16,390 ha), Athani (31,119 ha). Khanapur (9,640 ha), Chikodi (18,143 ha), Bailhongal (12,238 ha), Hukkeri (9,006 ha) and *Belgaum (6,059 ha) taluks irrigate area was 32.43, 21.24, 19.05, 18.45. 17.89,17.89, 17.07, 13.41, 13.38 and 9.32 percentage to their net sown area respectively. The above analysis reveals that the progress of irrigation in *Belgaum district is quite impressive and adequate to effectively meet the challenge of the high yielding technology. Though, the irrigated area is significantly improved in the district, it is necessary to study the different sources of irrigation to know the importance of particular mode of irrigation, which has been analysed below. Out of the total 1,14,575 ha of irrigated area, 44;618 ha (38.94%) were irrigated by canals, 10,760 ha (9.39%) by tanks, 43,476 ha (37.95%) by wells and 15,721 ha (13.72%) by other sources during 1974-75. But during 1984-85, out of total irrigated area of 1,99,392 ha, 79,633 ha (39.94%) were irrigated by canals, 8,770 ha (4.40%) by tanks, 57,850 ha (29.03%) by wells and 53,139 ha (26.63%) by other sources. (Table 43.1).

Ghataprabha Project

The Ghataprabha irrigation project has been undertaken to harness the waters of the Ghataprabha river to irrigate some semi-arid taluks of *Belgaum and Bijapur districts. In *Belgaum district, four taluks, i.e., Gokak, Raibag, Chikodi and Athani are irrigated under this project. It was estimated to irrigate nearly 69,521 ha of land in the year 1974-75 ha (61.19%). According to 1984-85 statistics 68,230 ha (85.76%) of land was irrigated out of estimated land of 79,557 ha, under this project (Table 43.2).

Malaprabha Project

This an important irrigation project undertaken by Government of Karnataka to irrigate the land of some semi-arid taluks of the districts of *Belgaum, Dharwad and Bijapur. This project has an estimated irrigable

Table 43.1 : *Belgaum District Area Irrigated by Different Sources

Area in hectares
Figures in brackets are %

TALUKAS	1974-75					1984-85				
	Net Area Irrigated (%) to net area sown)	*By Canals*	*By Tanks*	*By Wells Sources*	*By Other (% to net area sown)*	*Net Area Irrigated*	*By Canals*	*By Tanks Sources*	*By Wells*	*By Other*
Athani	19,514 (11.50)	518 (2.65)	398 (2.03)	14,508 (74.34)	4,090 (20.95)	31,119 (18.45)	1,243 (3.99)	811 (2.61)	9,938 (31.94)	19,127 (61.46)
Bailhongal	6,368 (7.0)	– –	2,964 (46.54)	1,831 (28.75)	1,573 (24.70)	12,238 (13.41)	– –	2,405 (19.65)	4,801 (39.23)	5,032 (41.15)
*Belgaum	5,912 (10.50)	334 (5.64)	2,375 (40.17)	2,210 (37.38)	993 (16.79)	6,059 (9.32)	– –	1,180 (18.48)	4,376 (72.22)	503 (8.30)
Chikodi	7,569 (8.60)	165 (2.17)	502 (6.63)	4,258 (56.25)	2,644 (34.93)	18,143 (17.07)	1,678 (9.25)	33 (0.18)	10,451 (57.60)	5,981 (32.97)
Gokak	27,727	19,576	–	6,710	1,441	37,367	28,415	–	5,178	3,774
Hukkeri	5,284 (7.90)	57 (1.07)	29 (0.54)	3,215 (60.84)	1,983 (37.52)	9,006 (13.38)	360 (4.00)	94 (1.04)	4,853 (50.89)	3,969 (44.07)
Khanapur	5,861 (11.70)	– –	3,831 (65.36)	355 (6.05)	1,675 (28.57)	9,640 (17.89)	– –	3,355 (34.80)	2,010 (20.85)	4,275 (44.35)

Raibag	30,855	22,282	283	7,938	352	35,994	21,247	–	11,378	3,369
	(51.90)	(72.21)	(0.91)	(25.72)	(1.14)	(51.35)	(59.03)	–	(31.61)	(9.36)
Ramdurg	3,300	1,010	135	1,300	855	16,390	12,229	137	2,712	1,312
	(3.60)	(30.60)	(4.09)	(39.39)	(25.90)	(19.05)	(74.61)	(0.84)	(16.55)	(8.00)
Soundatti	2,185	676	243	1,151	115	23,436	14,461	755	2,423	5,797
	(1.80)	(30.93)	(11.12)	(53.53)	(5.26)	(21.24)	(61.70)	(3.22)	(10.34)	(24.74)
District	1,14,575	44,618	10,760	43,476	15,721	1,99,392	79,633	8,770	57,850	53,139
Total	(12.70)	(38.94)	(9.39)	(37.95)	(13.72)	(21.34)	(39.94)	(4.40)	(29.03)	(26.63)

Source : Belgaum District : Talukawise Plan Statistics 1974–75 and 1984–85.

land of 47,832 ha under three taluks (Soundatti, Ramdurg and Bailhongal) of *Belgaum district, of this estimated land only 1,686 ha (3.52%) of land was irrigated in 1974-75. According to 1984-85 statistics 40,966 ha (82.70%) of land was irrigated out of estimated land of 53,163 ha. Soundatti taluk has major share, i.e., 24,874 ha of land under irrigation through this project (Table 43.2).

Table 43.2 : Irrigation by River Project - 1984-85 in Belgaum District

River Project	*Talukas*	*Area Estimated (in hectares)*		*Area Irrigated (in hectares)*	
Ghataprabha Project	Gokak	33,207	(41.75%)	30,645	(92.28%)
	Raibag	31,446	(39.55%)	29,842	(94.84%)
	Chikodi	6,988	(8.78%)	5,789	(82.84%)
	Athani	7,896	(9.92%)	1,963	(24.86%)
		(79.557)	(59.94%)	68,230	(24.86%)
Malaprabha project	Soundatti	30,128	(56.67%)	24,874	(82.56%)
	Ramdurg	14,929	(28.08%)	13.167	(88.20%)
	Bailhongal	8,106	(15.25%)	2,925	(36.08%)
		53,163	(40.06%)	40,966	(82.70%)
	GRAND TOTAL	1,32,720		1,09,196	(82.28%)

Source : Agricultural Production Programme, *Belgaum Dist. 1984-85.

Canal Irrigation (By Ghataprabha and Malaprabha Reservoirs)

Canal irrigation is the major irrigation system in the district. Table 43.1 reveals that 44,618 ha (38.94%) of area was under canal irrigation during the year 1974-75 which rose to 79,633 ha (39.94%) in 1984-85. The net increase of area under canal irrigation was 35,015 ha (78.48%). Looking at taluka wise area under canal irrigation, it was noticed that, Raibag and Gokak taluks had lions share with 22,282 ha (72.21%) and 19,576 ha (70.60%) respectively, to the irrigated area in the year 1974-75. This was followed by Ramdurg, 1,010 ha (30.60%); Soundatti, 676 ha (30.93%); *Belgaum, 334 ha

(5.64%) Athani, 518 ha (2.65%); Chilkodi, 165 ha (2.17%) and Hukkeri 57 ha (1.07%). There was no canal irrigation in the taluks of Bailhongal and Khanpur. During 1984-85, Gokak taluk shares highest land under canal irrigation, i.e., 28,415 ha (76.04%) followed by Raibag 21,247 ha (59.03%), Soundatti, 14,416 ha (61.70%), Ramdurg 12,229 ha (74.61%), Chikodi, 1,678 ha (9.25%), Athani, 1,243 ha (3.99%) and Hukkeri taluk, 360 ha (4.00%). No canal irrigation is noticed in Bailhongal, *Belgaum and Khanapur taluks till today.

Tank Irrigation

During 1974-75, 10,760 (9.39%) ha of land was under tank irrigation which reduced to 8,770 ha (4.40%) by 1984-85, which is mainly because of the development of canal irrigation in the district. Khanapur taluk had 3,831 ha of land under tank irrigation (65.36%) during 1974-75, and it was reduced to 3,355 ha during 1984-85 mainly because of the development of well irrigation and irrigation by other sources. Bailhongal and *Belgaum taluks had 2,964 ha and 2,375 ha of land respectively under tank irrigation during 1974-75. It reduced to 2,405 ha in Bailhongal and 1,180 ha in *Belgaum taluk during 1984-85. The remaining five taluks, i.e., Athani (398 ha), Chikodi (502 ha), Hukkeri (21 ha), Raibag (283 ha), Ramdurg (135 ha) and Soundatti (243 ha) had tank irrigation during 1974-75. No tank irrigation is seen in Gokak taluk during 1974-75.

Well Irrigation

Well irrigation also plays an important role in agricultural development of *Belgaum district. Very next to canal, well irrigation is the important source of irrigation of the district sharing 43,476 ha (37.95%) of land during 1974-75 and 57,850 ha (29.03%) of land during 1984-85. Since, it is an indigenous method of irrigation it is seen in most of the areas of the district. The water from the well is lifted by various methods like "kapali", "oil engines" and "electric motors" to irrigate the fields. Well irrigation varies from one taluk to another. During 1974-75, Athani taluk had the highest well irrigated area amongst all the taluks, 14,508 ha (74.35%) and it was followed by Hukkeri, 3,215 ha (60.84%), Chikodi, 4,258 ha (56.25%), Soundatti, 1,151 ha (53.53%), Ramdurg, 1,300 ha (39.39%), *Belgaum, 2,210 ha (37.38%), Bailhongal, 1,831 ha (28.75%), and Khanapur, 355 ha (6.05%). During 1984-85, Belgaum taluk shares the highest percentage, i.e., 72.33% (4,376 ha) of land under well irrigation and it was followed by Chikodi, 57.60% (10,451 ha), Hukkeri, 50.59% (4,583 ha), Bailhongal, 39.22% (4,801 ha), Athani, 31.94% (9,938 ha), Raibag, 31.63% (11,388 ha), Khanapur, 20.85% (2,010 ha), Ramdurg, 16.55% (2,712 ha), Gokak, 13.86% (5,178 ha) and Soundatti, 10.34% (2,423 ha). Well irrigation is the best source to the poor and marginal farmers to irrigate their limited area of land though it is costlier in the initial stage. Except Athani and Gokak taluks remaining all eight taluks increased their land under well irrigation during 1984-85, compared to 1974-75 statistics.

Irrigation by Other Sources

In addition to canal, well and tank irrigation, the other sources like lift irrigation are also used to irrigate agricultural land. During 1974-75, *Belgaum district had 15,721 ha (13.72%) of land under irrigation by other sources. Athani taluk had highest arable land (4,090 ha) under irrigation by other sources. Chikodi taluk rank second with 2,644 ha of land under irrigation by other sources. Remaining eight taluks had their land under irrigation by other sources ranging from 115 ha to 1,983 ha. During 1984-85, the district had 53,139 ha (2.663%) of land under irrigation by other sources. During this year Athani taluk again shares highest arable land (19,127 ha) under irrigation by other sources. Bailhongal (5,032 ha), Chikodi (5,981 ha) and Soundatti (5,797 ha) taluks had more or less equal land under irrigation by other sources. Remaining six taluks had land under irrigation by other sources, ranging from 503 ha to 4,275 ha. In the district, during a span of 10 years, 37,418 ha (238.01%) of land was increased under irrigation by other sources. In *Belgaum district during 1984-85 land under irrigation by other sources has been increased to 238.01% when compared to 1974-75. This significant increase is mainly due to extension of financial assistance to small and marginal farmers. Except *Belgaum taluk almost all taluks have raised more land under this category of irrigation in the district

Intensity of Irrigation

Intensity of irrigation is controlled by various factors such as sources of irrigation, types of crops grown, cropping season, quantity and quality of water supply, density of network of water channels, etc. The benefits of intensive irrigation are reflected in the cropping pattern, productivity of land, land use efficiency and method of cultivation. In an agricultural region, other thing being equal, the intensity of irrigation increase with decrease of rainfall and vice versa (Jasbir Singh and Dhillon, 1984). The intensity of irrigation will always remain low and negligible in rain fed areas where there is restricted surface water. Limited salt-ridden sub-soil water and hilly or undulating topography. The intensity of irrigation is not uniform in *Belgaum district. The district as a whole, intensity value was 103.30 in 1974-75 whereas it was 121.83 in 1984-85. The net increase in the intensity of irrigation was 18.53 (Table 43.3, Fig. 43.1).

Table 43.3 : Intensity Categories of Irrigation for 1974–75 and 1984–85 for *Belgaum district

1974–75				1984-85			
Intensity Categories	*Range of Intensity*	*No. of Talukas*	*Name of the Talukas*	*Intensity Categories*	*Range of Intensity*	*No. of Talukas*	*Name of Talukas*
1	2	3	4	1	2	3	4
Very Low	Below–99.42	–	–	Very Low	Below–105.04	4	Athani, Hukkeri, Khanapur
Low	99.43–103.38	7	Athani, Bailhongal, Chikodi, Hukkeri, Khanapur, Raibag and Soundatti.	Low	105.05–118.06	3	*Belgaum, Chikodi, Ramdurg.
Medium	103.39–107.34	2	Belgaum and Gokak	Medium	118.07–131.08	3	Bailhongal, Raibag, Soundatti.
High	107.35–111.30	–	–	High	131.09–144.10	1	Gokak.
Very High	Above–111.30	1	Ramdurg	Very High	Above–144.10	–	–

The taluk level analyses reveals that during 1974-75, the intensity of irrigation was very high in only one taluk, i.e., Ramdurg (113.96). The low intensity was in Athani, Bailhongal, Chikodi, Hukkeri, Khanapur, Raibag and Soundatti taluks (between 99.43 and 103.38), medium intensity was in the taluks of *Belgaum and Gokak (between 103.39-107.34). There were no taluks in high and very low intensity of irrigation during 1974-75. The use of river project water was low as a result intensity of irrigation in the district was also low.

In 1984-85, the intensity of irrigation increased substantially due to completion of river projects. The high intensity was noticed in Gokak taluk (141.76) and the medium intensity range was observed in Bailhongal, Raibag and Soundatti taluks (118.07-131.08). *Belgaum, Chikodi and Ramdurg taluks were in low intensity range (105.05-118.06), whereas, Athani. Hukkeri and Khanapur taluks were in very low intensity range (below 105.04). Almost all taluks increased their intensity of irrigation over a period of ten years, i.e., 1974-75 to 1984-85.

Potentiality of Surface and Groundwater

According to the report of the Department of Mines and Geology, Government of Karnataka, Karnataka has nearly 1.5 lakh million cubic metres (mcum) of potential water for future use. This potential varies from one district to another. Chikamagalore and Coorg districts stand first in total water potentiality, while *Belgaum and other districts stand in the second order. *Belgaum district is utilising about 56.91 mcum of water for irrigation (Table 43.4). Therefore, out of 12,834.27 mcum of total water resources of the district, still nearly 12,800 mcum of water is to be utilised for other uses.

Table 43.4 : Distribution of water resources of different uses in Belgaum District

A	*B*	*C*	*D*	*E*	*F*	*G*	*H*	*I*
12,834.27	0.01	56.91	30.17	1.91	–	4,200.90	2.52	8,541.85

Note :

A: Total water resources in mcum.

B: Per capita water resources availability in mcum.

C: Water used for irrigation in mcum.

D: Drinking and domestic water requirement in mcum.

E: Water used in the industries in mcum.

F: Water used for power generation in mcum.

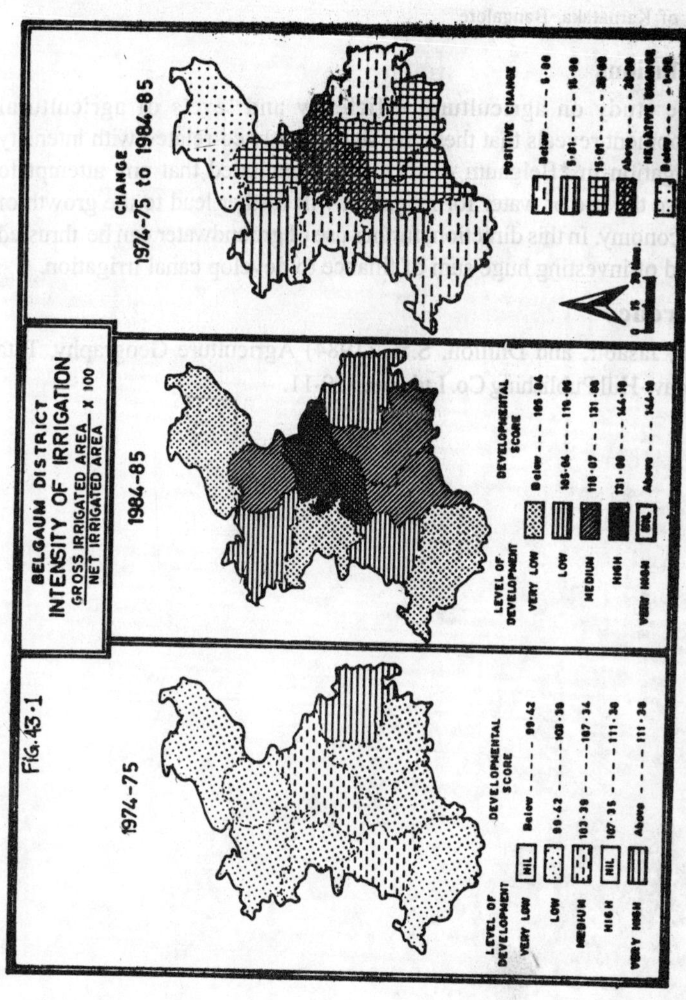
FIG. 43·1
BELGAUM DISTRICT
INTENSITY OF IRRIGATION
GROSS IRRIGATED AREA / NET IRRIGATED AREA X 100
1974-75
LEVEL OF DEVELOPMENT
DEVELOPMENTAL SCORE
VERY LOW
NIL
Below ----- 99·42
LOW
MEDIUM
HIGH
NIL
107·35 ---- 111·38
VERY HIGH
Above ---- 111·38
1984-85
LEVEL OF DEVELOPMENT
DEVELOPMENTAL SCORE
VERY LOW
Below --- 105·04
LOW
MEDIUM
HIGH
VERY HIGH
NIL
CHANGE
1974-75 to 1984-85
POSITIVE CHANGE
Below
Above
NEGATIVE CHANGE
Below
0 15 30 Kms.

G: Annual evaporation in mcum.

H: River run-off in mcum.

I: Potential water resources in mcum.

(Source: Ground water studies, Department of Mines and Geology, Government of Karnataka, Bangalore.

Conclusions

The study on agricultural efficiency and levels of agricultural development reveals that these are significantly co-related with intensity of irrigation in *Belgaum district. It is concluded that any attempt to increase the use of water for irrigation will further lead to the growth of agroeconomy. In this direction, further use of groundwater can be thrusted instead of investing huge sum of finance to develop canal irrigation.

Reference

Singh, Jasabir, and Dhillon, S.S., (1984) Agriculture Geography, Tata McGraw-Hill Publishing Co. Ltd., pp. 110-11.

44

Water Resources Development and Management in India

K. NARASIMHA MURTHY

Introduction

Natural resources are the greatest asset of any country. It is well known fact that water is a scarce commodity and is not spread over according to the needs of the place or the person. Thus a systematic planning of the conservative and judicious use of the water resources is very important for the development of any region. Before considering the problems and prospectus of further development of water resources, a survey of available resources and their state of development would be useful. This article is an overview of the present availability and requirements for future need.

India with a geographical area of 329 million hectares and a population of 658 millions (1981) is blessed with large river systems. It possesses about 4% of the total average annually runoff in the rivers of the world. However, the per capita water availability of natural runoff is only 2,500 cubic metres (cum) (1987) per year compared to the availability of more than 1,700 cum in USSR, 6,500 cum in Japan and 6,200 cum in USA. With the present rate of growth of population in our country the figure 2,500 cum may dwindled down further by turn of century. The problems posed are formidable, but nature's bounty to our country is such that we can surmount these problems by proper natural resource management.

Water Resources

(a) Climate

The Indian climate is basically tropical and monsoonic. Except in the mountains, the temperatures remain well above freezing throughout the year permitting year round cultivation. There is ample sunshine all through the year.

The average rainfall over the entire country is about 120 cm. (400 million hectare metres). There is a large spatial and temporal variation. Cherapunji with an annual rainfall of 1,142 cm. is the wettest place on earth whereas the Thar desert is one of the most arid regions in the world (Fig. 44.1). Most of the rainfall amounting to 70% to 80% is concentrated during the

FIG. 44·1
ANNUAL RAINFALL

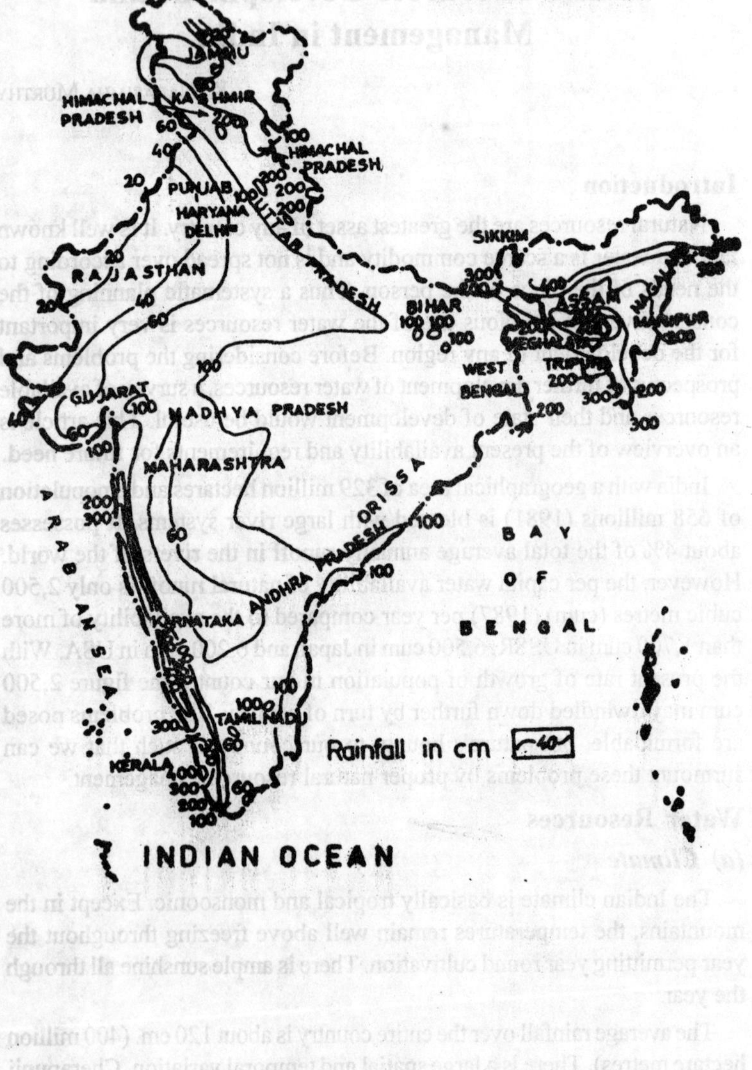

Rainfall in cm

four monsoon months. The annual potential evapotranspiration varies from 140 to 250 cm. over most part of the country. If runoff is not stored, it not only goes waste to sea but also cause flood damage. Irrigation is an essential requirement for productive agricultural activity in the country.

(B) Assessment

Assessment of water resources of India has been attempted at different agencies in our country (Iyer, 1989). The Central Water Commission arrived at a value of 1,880 Km^3 as the total average annual water potential of the country. This figure is a combined estimate for surface and groundwater. The basin wise annual average water potential of the country is given in Table 44.1 (Iyer,1989). Owing to the topographical and other constraint, only 1,110 km, of available water can be put to beneficial use. The possible utilisation of groundwater as estimated by Central Ground Water Board is 420 km^3/year. The basinwise break up of the utilisable flows is given in Table 44.2 (Iyer, 1989).

Development of Water Resources

(a) Past Scenario

Efforts have been made in India since times immemorial, consistent with the techno-economic capability of the people, to develop tanks, wells and small canals for the irrigation purposes. From legendary accounts down to latest planning documents water resources utilisation has been considered to be of greatest importance in India. This is understandable in view of its crucial importance for improving the quality of life in the agrarian economy and the peculiar climatic and hydrological conditions, yet till as late as the beginning of the sixth five year plan, only about 10 per cent of the rural population could be supplied with safe drinking water and only 30 per cent of the cultivated land could be irrigated.

Many important irrigation works, such as Western Yamuna Canal in 1820, Eastern Yamuna and Upper Ganga canal during 1830-54, the Upper Bari Doab Canal in 1851, Godavari and Krishna anticuts in 1850, Agra canal in 1873, to name a few were constructed as famine relief works. The First Famine Commission of 1880 emphasised the need for direct state initiative in the development of irrigation, particularly in the vulnerable areas. No comprehensive plan of irrigation was, however, formulated presumably because of the complacency engendered by the comparatively good agricultural years from 1880 to 1895. It was again the tragedy of famines in 1897-98 and 1899-1900 that led to setting up the First Irrigation Commission in 1901 to ascertain the utility of irrigation as a means of protection against famine, the extent of irrigation development and the

Table 44.1 : Basinwise Average Annual Water Resources (Km³)

Sl. No.	*Basin*	*Average annual Run off in the river*	*Annual Ground Water draft 1983-84*	*Average annual additional evapotrans-piration attributable to groundwater use*	*Total av. Annual Potential*
(1)	(2)	(3)	(4)	(5)	(6)
1.	Indus (Up to border)	73.305	12.79	Negligible	73.305
2.	(*a*) Ganga	501.643	49.75	23.380	525.023
	(*b*) Brahmaputra (at Jogigula)	537.067	0.34	0.173	537.24
	(*c*) Barak & other rivers flowing into Meghana like Gumti, etc.	59.800	0.01	Negligible	59.800
3.	Godavari	118.982	5.99	Negligible	118.982
4.	Krishna	67.790	4.92	Negligible	67.790
5.	Cauvery	20.957	3.34	0.301	21.358
6.	Pennar	6.858	1.27	Nil	6.858
7.	East Flowing rivers between Mahanadi & Pennar	16.948	1.47	Nil	16.948
8.	East Flowing rivers between Pennar & Kanyakumari	17.725	7.53	Nil	17.725
9.	Mahanadi	66.879	0.75	Negligible	66.879
10.	Brahmani & Baitarani	36.227	0.45	Nil	36.227
11.	Sabarnarekha	10.756	0.18	0.038	10.794

Table 44.1 (Contd.)

(1)	(2)	(3)	(4)	(5)	(6)
12.	Sabarmati	3.812	1.27	0.267	4.079
13.	Mahi	11.829	1.03	Nil	11.829
14.	West flowing rivers of Kutch, Saurashtra including Luni				
		15.098	3.88	Nil	5.096
15.	Narmada	40.950	1.29	0.323	41.273
16.	Tapi	18.00	1.69	0.389	18.389
17.	West flowing rivers from Tapi to Tadri				
		108.618	0.74	0.392	109.010
18.	West flowing rivers from Tadri to Kanyakumari				
		89.250	1.14	0.564	89.844
19.	Area of Inland drainage in Rajasthan desert				
		–	0.22	–	–
20.	Minor river basins draining to Bangladesh & Burma				
		31.000	0.005	Negligible	31.000
	Total	1,850.000		25.927	1,880.00

scope for further irrigation works. About this time the gross area irrigated from public works in British India was about 7.5 million hectares (mha) of which 3 mha was from minor works like tanks, inundation canals, etc. The area irrigated by protective works was only a little over 0.12 mha. Private works irrigated 5.7 mha about 70 per cent by wells and the remaining from tanks, streams and channels, etc. The total gross irrigated area was 13.3 mha. The first large scale groundwater development project envisaging construction of about 1,500 tube wells was launched in the country during mid 1930 in Ganga basin. Around 1945, the area irrigated from private sources in undivided India, excluding princely states, was 10 mha of which 4.4 mha was from wells. The progress of development of irrigation of British in the current century prior to independence is shown in Table 44.3.

Table 44.2 : Basinwise Annual utilisable Water Resources (Km³)

Sl. No.	Basin	Estimated utilisable flow excluding ground water	Utilisable groundwater (CGWB assessment 1983-84)	Total utilisable flow
1.	Indus (Up to border)	46.000	17.810	63.810
2.	(a) Ganga	250.000	172.010	422.010
	(b) Brahmaputra	24.000	20.820	46.150
	(c) Barak	–	1.330	–
3.	Godavari	76.300	44.980	121.280
4.	Krishna	58.000	24.620	82.620
5.	Cauvery	19.000	10.420	29.420
6.	Pennar	6.858	5.350	12.208
7.	East flowing rivers between Mahanadi & Pennar	13.110	11.690	24.800
8.	East flowing rivers between Pennar & Kanyakumari	16.732	21.080	37.812
9.	Mahanadi	49.990	18.200	68.190
10.	Brahami & Baitarni	18.297	7.890	26.187
11.	Subarnarekha	6.813	2.850	9.663
12.	Sabarmati	1.925	4.380	6.305
13.	Mahi	3.095	4.440	7.535
14.	West flowing rivers of Kutch, Saurashtra including Luni	14.980	12.610	27.590
15.	Narmada	34.500	13.000	47.500
16.	Tapi	14.500	6.730	21.230
17.	West flowing rivers from Tapi to Tadri (including Tadri)	11.936	8.890	20.916
18.	West flowing rivers from Tadri to Kanyakumari	24.273	7.740	32.013

Table 44.2 (Contd.)

Sl. No.	Basin	Estimated utilisable flow excluding ground water	Utilisable groundwater (CGWB assessment 1983-84)	Total utilisable flow
19.	Area of Inland drainage in Rajasthan desert	–	1.330	1.330
20.	Minor river basins draining to Bangladesh & Burma	–	0.280	0.280
	Total	690.00	418.540	1,110.00

The socio-political objective of irrigation was thus only to stabilize the sustenance agriculture for which extensive irrigation became the logical policy. Technological, economic and financial considerations also supported this policy. The colonial government was content with supplying the water up to canal outlets, leaving the rest of the task of conveyance and distribution to the farmers themselves.

The issue of activities such as land levelling, drainage, lining of water courses, etc., which demanded public participation and was not consistent with the objectives of a foreign government and was not within the capability of the poor peasantry. Thus, in short, a policy of extensive irrigation through low flow diversions to stabilize the sustenance agriculture evolved from a convergence of social, political, economic, technological and financial considerations.

Table 44.3 : Progress of Irrigation Development in Undivided India (Singh, 1987)

(Net irrigated area in million hectares)

Year	Public Sector	Private Sector	Total
1990	7.6	5.7	13.3
1920-21	10.4	8.9	19.3
1945	13.5	10.0	23.5

After independence in 1947, India was left with 82% of the population of undivided India and 84% of net land area but only 69% of irrigated area. Consequently, several large river development projects such as Bhakra Nangal, Damodar Valley and Hirakud schemes were taken up between 1947 and the start of the first five year plan. High priority was attached to irrigation development in the subsequent plan periods. This may be observed from the Tables 44.4 and 44.5.

It is quite evident from the Tables 44.4 and 44.5 below the higher outlays than before were spent in every subsequent plan period both on major, medium and minor irrigation schemes. In terms of percentage, the expenditure incurred on major and medium irrigation projects alone constituted 19.5 per cent of total plan expenditure in the first five year plan. The subsequent plans saw a decline in this percentage. Even then it has always been between 6 to 9 per cent.

More or less similar trend is noticeable for minor irrigation schemes. It would be noticed from Table 44.5 that a significant role has been played by the financial institutions of the development of minor irrigation. Hence, from a negligible start in the first five year plan the share of institutional

Table 44.4 : Development of Major and Medium Irrigation Projects Since 1951 (Sharad Kumar, 1986)

Period	*Expenditure (Rs. Crores)*	*Expenditure as % of total plan expenditure*	*Irrigation Potential Created (X 10^6 ha)*
Pre-plan	–	–	9.70
I Plan	380	19.5	2.49
II Plan	380	8.1	2.14
III Plan	581	6.8	2.25
Annual Plan	434	6.6	1.54
IV Plan	1,337	7.8	2.60
V Plan	2,442	6.2	4.12
Annual Plan (1978-80)	2,056	8.9	1.79
VI Plan	7,510	8.8	4.35
Total	–	–	30.85

Table 44.5 : Development of Minor Irrigation Scheme Since 1951 (Sharad Kumar, 1986)

Period	*Plan expenditure (Rs. Crores)*	*Institu-tional quality*	*Potential Created (X 10⁶ ha)* Minor Surface irrigation	Ground-water irrigation	*Total*
Pre-plan	–	–	6.4	6.5	12.90
I Plan	76	Negligible	0.0	1.1	1.16
II Plan	142	19	0.1	0.7	0.73
III Plan	326	115	0.0	2.2	2.21
Annual Plan	326	235	0.0	2.0	2.00
IV Plan	513	661	0.5	4.0	4.50
V Plan	613	780	1.0	4.1	5.10
Annual Plan (1978-80)	497	490	–	–	1.40
VI Plan	1979	1,437	–	–	7.00
VII Plan	3228	3,311	–	–	–
Total	–	–	–	–	37.00

outlay has gone up to Rs. 1,700 crores which is almost equal to the planned expenditure of Rs. 1,810 crores. As a matter of fact institutional outlay exceeded the planned expenditure during the fourth and fifth plans.

After independence, although highest priority has been laid on irrigation development and considerable development has indeed taken place, it has continued on technologically oriented adhoc project by project basis without any integrated policy analysis and in fact the colonial policy has continued to be followed uncritically. A high technology bias is also distinctly evident. Even higher dams and larger canals have been the sole preoccupation of all concerned. Projects continued to be taken up on an adhoc basis without detailed coordinated planning. Completion periods of projects get extended on various grounds and irrigation adaptation follows a leisurely pace. Conceptual awareness that integrated river basin planning and conjunctive surface and groundwater irrigation development should be undertaken, are often stated in government announcements or commission's reports, but neither these are planned nor implemented.

Indeed a policy analysis was undertaken only by the Second Irrigation Commission Report, (1972) is more of an assemblage of facts on an adhoc enunciation of some policy issues, but it totally lacks the critical analysis needed for scientific formulation on water resources development policy.

(b) Present Scenario

Irrigation as at present practised in the country is somewhat extravagant in the use of water. On an average about 0.8 ha-m of water is required to irrigate a hectare. On this basis, the available water resources would suffice for irrigating about 96 mha. On many irrigation systems the present mode of utilisation of water is wasteful. On unlined canals in the alluvial tracts only about two-fifths of water released at the canal head is utilised by crops; the rest is lost in transit and in the fields. Not all of these losses can be eliminated, but obviously there is a scope for saving water. For this, lining of channels, proper levelling of fields and adoption of most suitable mode of application of water to crops are important steps. The area which can be irrigated with a given quantity of water also depends on the cropping pattern. It is envisaged that with more scientific and economical use of water in the future, the average water depth for irrigation will be substantially reduced. In view of the inadequacy of water resources to meet the requirement, there is need for most emphasis on efficiency and economy in water use.

Much of the water available in irrigation schemes is underutilised for lack of attention to land levelling, field distribution, drainage system and regulatory services. Because water is ineffectively used, crop yields are much less than they could be.

Although percentage of utilisation for purposes other than agriculture is low at present, this is expected to rise appreciably in the future with increase in industrialisation and power generation through thermal and nuclear plants. Part of these requirements will have to be met with sea water by locating some of the industries and power plants, requiring large amounts of water for cooling on the sea coasts. On rivers with large unutilised flows like the Ganga, more water should be used by diverting the flood flow for irrigating crops during kharif season which would also increase groundwater recharge. Also, greater recharge of aquifers can be secured along rivers and streams by depressing the pre-monsoon groundwater table through a larger exploitation of groundwater. Wherever feasible, surplus water in a river, such as in the Brahmaputra, the Ganga and west flowing rivers South of Tapi, should be utilised. All these possibilities need to be examined and considered.

(c) Future Scenario

Nearly 85% of the present use has been for agriculture. Domestic and industrial water needs have largely been concentrated in or near principal cities. Table 44.6 indicates the position of utilisation during 1985, the projected demands of water for various purposes as worked out by National Commission on Agriculture (NCA) (1976) and updated with the available data (Iyer, 1989). The growth process, the increase in population and the expansion of economic activities inevitably lead to increasing demand for water, which is estimated to be 1,050 km^3 by the year A.D. 2025. Hence, overall national planning and resources management in respect of water is necessary. There is need for co-ordination and allocation of priorities among the diverse uses. The ultimate irrigation potential from major, medium and minor irrigation schemes is estimated at 113 mha of which 58 mha is from major and medium schemes and 55 mha from minor irrigation schemes.

Hydropower Development

The utilisable hydropower potential of the country is estimated as about 75,000 MW. Unfortunately, the rate of growth of hydropower in the country has been slow and uneven. The per cent hydropower increased from 24 in 1950 to a peak to 41.4 in 1969. After that the stress was shifted to thermal till 1980. Out of the total installed capacity of 31,024 MW only 11,383 or 32.9% was hydropower. Practically the same ratio continued till 1987. The ostensible reason pleaded for this was the higher capital investment and longer construction time of hydropower plants.

Table 44.6 : Annual Requirement of Fresh Water (km^3) (Iyer, 1989)

	1985 A.D.		*2000 A.D.*		*2025 A.D.*	
	Surface water	*Ground water*	*Surface water*	*Ground water*	*Surface water*	*Ground water*
1. Irrigation	320	150	420	210	510	260
2. Other uses	40	30	80	40	190	90
Total	360	180	500	250	700	350

Structure of Irrigation Development

The irrigation potential has increased from 22.6 mha in 1950-51 (pre-plan period) to 67.85 mha in 1984-85. This increase in irrigation potential has been made possible through both means of surface and groundwater.

Between 1950-51 and 1984-85 groundwater irrigation has been doubled. This groundwater is today responsible for nearly half in terms of area and much more than half in terms of irrigation efficiency and productivity (Vohra, 1980).

The advent of green revolution ushered a new era in the programme of groundwater development leading to large scale construction of dug wells, private shallow tube wells and public tube wells (Table 44.7). Other facts which accelerated the pace of groundwater development during the past three decades are the flow or institutional finance and groundwater subsidies for its development and rural electrification.

It is observed that the share of minor irrigation was more than major and medium irrigation at 13.1 mha in the pre-plan period compared to 9.7 mha. By 1984-85 the cumulative total of irrigation potential created through minor irrigation was around 37 mha compared to around 30.85 mha of major and medium irrigation projects.

It is found that the utilisation of irrigation potential created in the pre-plan period (1950-51) stood at 100 per cent for both major, medium and minor projects. During 1979-80, however, it was 100 per cent utilisation for minor irrigation (both surface and groundwater) and only 85 per cent for major and medium projects (Table 44.8).

Table 44.7 : Growth of Groundwater Structures (Dutta, 1989)

	Groundwater potential developed (mham)	*Dug wells (000 nos.)*	*Shallow Tube wells (000 nos.)*	*Public Tube well*
1950-51	6.5	3,860	3	2,400
1960-61	8.3	4,540	22	8,900
1968-69	12.5	6,110	360	14,700
1973-74	16.6	6,700	1,140	22,000
1979-80	22.0	7,786	2,132	33,300
1984-85	27.8	8,742	3,360	46,166
1989-90 (anticipated)	35.6	9,990	4,778	71,136

Table 44.8 : Utilisation of Irrigation Potential (Sharad Kumar, 1986)

	1950-51		*1960-61*		*1968-69*		*1979-80*	
	P	*U*	*P*	*U*	*P*	*U*	*P*	*U*
1. Surface Water (a + b)	16.1	16.1	21.0	19.7	25.1	24.0	35.6	30.6
(*a*) Major & Medium	9.7	9.7	14.4	13.1	18.1	17.0	26.6	22.6
(*b*) Minor	6.4	6.4	6.6	6.6	7.0	7.0	8.0	8.0
2. Groundwater	6.5	6.5	8.2	8.2	12.0	12.0	22.0	22.0
Total	22.6	22.6	29.2	27.9	37.1	36.0	57.6	52.6

P – Potential and U – Utilisation, in mham

Compared with industrial statistics these figures of utilisation appear very impressive. It is therefore necessary to point out that these figures relate rarely to the supply of water and say very little of productivity either per unit of land or per unit of water. Even the minor irrigation projects appear to have performed better than major and medium projects. The situation in regard to major and medium irrigation projects, seems to be deteriorating over the years.

It is therefore relevant to argue that there are measures of efficiency other than utilisation capacity like which of the two projects is similar to manage?' and 'which of the two is more productive from the yield point of view?' etc. But equally important is the problems of waterlogging and salinity. It is therefore, of significance to note that area threatened in India by water-logging and salinity in irrigated area has been estimated to be around 10 mha. This constitutes almost 20 per cent of the total irrigated area in the country and more than 30 per cent of irrigated area under major and medium irrigation projects. The adverse impact of water-logging, salinity and seepage can be reduced if the canals of the projects are lined.

Another way to face this menace of water-logging is the conjunctive use of surface and groundwater (Ex : Haryana and Punjab). In this regard it is observed that the overall advantages to the country through integrated and conjunctive use are economically so attractive that the financial agencies need to be encouraged so that more and more farms take advantage of loan facilities (Ministry of Agriculture and Irrigation, 1976). Taking stock of Indian agriculture it is also observed, 'a very large proportion of the holdings in the country is less than a hectare and it may be difficult for an individual farmer with such a small holding to have his own private tube well. In such cases, a cooperative approach of grouping small holdings in adjacent areas and establishing irrigator's association or water user's cooperatives for developing groundwater is recommended' (NCA-Ministry of Agricultural and Irrigation, 1976).

Progress of Irrigation in Different States

Taking stock of the ultimate irrigation potential of each state the following picture emereges (Table 44.9) (Sharad Kumar, 1986). U.P., Bihar, Madhya Pradesh, Andhra Pradesh and Maharashtra ranked first, second, third, fourth and fifth states in respect of ultimate irrigation potential. Punjab even being a small state appears to be well endowed in irrigation potential and ranked sixth amongst all states. Moreover, Punjab, Tamil Nadu, U.P., Rajasthan and Haryana have harnessed the irrigation potential to a optimum level, thus narrowing the difference between the created irrigation potential and the ultimate potential. On the other hand states like Bihar (50%), M.P. (37%), Orissa (42%) and Assam (25%) have lagged behind.

Similarly, if we review the progress of irrigation in each state on the basis of : (*a*) major and medium irrigation and (*b*) minor irrigation the following picture emerges. The difference between created potential and ultimate potential has been narrowed in states like Tamil Nadu, Punjab, West Bengal and Andhra Pradesh which have reached an irrigation potential of 84 per cent, 82 per cent , 69 per cent and 68 per cent respectively (Table 44.9). It is equally interesting to note that state like Haryana does not have any perennial river flowing through the state, yet it can claim an irrigation potential of 3 mha from major and medium irrigation projects.

The performance of states like M.P., Bihar, Orissa, Maharashtra, Gujarat, Assam and Manipur is very dismal since they have not reached even 50 per cent of their ultimate potential from major and medium irrigation projects. Even the created irrigation potential in these states is not fully utilised. Barring Orissa (100%), all the remaining states stood below India's average capacity of utilisation (89 per cent).

Perhaps more significant is the performance of each state with regard to minor irrigation. The states, U.P. (93%), Haryana (91%), Gujarat (91%), Punjab (87%), Tamil Nadu (84%) and Rajasthan (83%) are narrowing down the gap between their created potential and the ultimate potential to a great extent. The performance of states like Assam (23%), Kerala (36%), Orissa (39%), West Bengal (44%) and M.P. (47%) is far from satisfactory.

Table 44.9 : Area Irrigated and the Ultimate Irrigation Potential ('000ha) (Sharad Kumar, 1986)

	Ultimate irrigation potential			*Area covered*			*Created potential percentage of ultimate irrigation potential*		
	Major & medium	*Minor*	*Total*	*Major & medium*	*Minor*	*Total*	*Major*	*Minor*	*Total*
U.P.	12,500	13,300	25,700	6,813	12,301	19,114	55	93	74
Bihar	6,500	5,900	12,400	2,879	3,282	6,161	44	56	50
M.P.	6,000	4,200	10,200	1,799	1,970	3,769	30	47	37
A.P.	5,000	4,200	9,200	3,391	2,260	5,657	68	54	61
Maharashtra	4,100	3,200	7,300	1,772	1,936	3,708	43	61	51
Punjab	3,000	3,550	6,550	2,467	3,097	5,564	82	87	85
W. Bengal	2,310	3,800	6,110	1,587	1,679	3,266	69	44	54
Orissa	3,600	2,300	5,900	1,572	900	2,472	44	39	42
Rajasthan	2,750	2,400	5,150	1,821	1,979	3,800	66	83	74
Gujarat	3,000	1,750	4,750	1,324	1,590	2,914	44	91	61
Karnataka	2,500	2,100	4,600	1,285	1,123	2,408	52	54	52
Haryana	3,000	1,550	4,550	1,927	1,415	3,342	64	91	74
Tamil Nadu	1,500	2,400	3,900	1,252	2,038	3,290	84	85	84
Kerala	1,000	1,100	2,100	578	390	968	58	36	
Assam	370	1,700	2,070	131	391	522	14	23	25
Jammu & Kashmir	250	550	800	144	336	480	58	61	60
Himachal Pradesh	50	285	335	6	118	124	12	41	37
Manipur	135	105	240	43	39	82	32	37	34
Tripura	100	115	215		46	46		40	21
Meghalaya	20	100	120		39	33		39	28
Nagaland	10	80	80		56	56		70	62
Sikkim	20	22	42		13	13		59	31

Jammu & Kashmir and Maharashtra (61%), Sikkim (59%), Bihar (56%), Andhra Pradesh (54%) and Karnataka (54%) have a scope for 100% increase in area under minor irrigation.

Water Resources Management

Water resources management essentially means proper manipulation of the hydrologic cycle. But the hydrologic cycle is great influenced by the various factors like the presence of forests, the urbanization which interferes with the natural soaking of water or which accentuates the accumulation of the surplus rain from the covered areas, the rail/road network which obstructs the sheet flows, diverts water and modifies the time of surface flows. Assessment of the impact of these factors on the different phases of the hydrologic cycle is therefore going to be very important for any water resources management.

The water resources development programmes need not necessarily be in conflict with the environmental requirements. In this context that the practice being adopted by some of our dam builders for diverting the entire available water away from the river without releasing anything back into its natural course needs a serious reconsideration. Total diversion of all the available supply throws the entire life supporting system along the river bank out of gear. No project should ever aim to totally dry up the river bed but should provide minimum flows, if possible perennially, down the river. Our effort should be to correct the imbalanced in nature and not to aggravate it.

Suggestions

(1) Realizing that the available waters are finite, and that there are limits to acquisition and exploitation of natural resources like water, it is time that we switch over from the accumulative or extractive technology to conservation oriented technology. With the latter type of orientation, there will be greater stress on utilising the limited resources in a more efficient manner rather trying to only acquire or extract more of the water from nature.

(2) The process of evaporation and the loss of water by capillary action from ground surface needs to be fully investigated and these losses should be checked.

(3) The best integrated development of the water resources is through construction of reservoirs across the river in a cascade pattern, i.e., in close succession one below the other.

(4) Basin-wise management of a river is essential. Also treatment of industrial effluents is necessary to check the river pollution.

References

Chaturvedi, A.C., (1987) Industrial requirement of water in the 1990's Roc. Symp. Integrated Water Resources Management for Drinking, Agriculture and Industry. Policies and Issues, IWRS, Hyderabad. pp. 113-21.

Dhawan, B.D., (1982) Development of tube well irrigation in India. Agricole Publishing Academy, New Delhi.

Dutta, D.K., (1989) Groundwater development. J. IWRS, V. 9, No. 2, pp. 60-62.

Iyer, S.S., (1989) Water resources of India. An overview. J. IWRS, V. 9, No. 2, pp. 7-11.

Ministry of Agriculture and Irrigation, Government of India, (1976) Report of expert committee on integrated development of surface and groundwater.

Madhavan, K., (1986) A reliable system for water management. Journal of Irri. and Power, pp. 347-48.

Mistry, J.F., (1987) Importance of irrigation development in India, J. IWRS, V. 7, No. 1.

Pant Niranjan, (1984) Community tube well organisational alternative to small farmers irrigation. Economic and Political Weekly, V. XIX, No. 26.

Patel, C.C., (1989) Water availability and use. IE(I) Bull. V. 39, pp. 1-9.

Sharad Kumar, (1986) Irrigation development in India–A review. WAMANA, pp. 1-8.

Singh Ranvir, (1987) Agriculture and water resources development in India–a historical perspective. Proc. Symp. Integrated Water Resources Management for Drinking. Agriculture and Industry, Policies and Issues, IWRS, Hyderabad, pp. 13-20.

Tripathy, D., (1987) Water and welfare goals conflicts and reconciliation. J. IWRS. V. 7, No. 4, pp. 44-52.

Vohra, B.B., (1980) A policy for land and water. Patel memorial lectures, New Delhi.

45

Water Resources of Bihar : Potentialities for Fishery Development

S. HASIB AHMAD
AND
ARUN K. SINGH

Introduction

Bihar is entirely a landlocked State of India and extends approximately from lat 22°N to 27°31'N. and 83°20'E to 88°18'E. The maximum north-south extent of Bihar is about 605 km. and the maximum east-west width is about 483 km. and occupies an area of about 1,74,000 sq. kilometres. The natural resources of Bihar include the area and extent of a territory, its surface and underground water resources, in land and soils, etc. In this paper an attempt has been made to study the water resources in relation to their potentialities and utilization for fish as well as other aquatic production which are of economic value and utilized as food for human beings and cattles. The aquatic resources comprise only the freshwaters. The various natural and man-made water resources of the state are categorised as follows :

I. Lentic (Still) Water Resources
- (*i*) Ponds and Tanks
- (*ii*) 'Mans' (Ox-bow lakes)
- (*iii*) 'Chaurs' (Floodplains)
- (*iv*) Reservoirs (Man-made Lakes)

II. Lotic (Running) Water Resources
- (*i*) Rivers
- (*ii*) Irrigation Canals

I. Lentic (Still) Water Resources

(i) Ponds and Tanks

There are myriads of ponds and tanks distributed throughout the length and breadth of the State (Table 45.1). Ponds and tanks are an essential feature of the rural setting of Bihar. Excavated in the ancient times mainly for community bathing, washing of cloths, cattle wallowing, irrigation and

Table 45.1 : Details of Ponds and Tanks of Bihar (Bihar State Planning Commission, 1990)

	Name of the District	*Water area in hectare*
1.	Aurangabad	3,511.38
2.	Begusarai	1,033.96
3.	Bhagalpur	918.00
4.	Bhojpur	1,096.20
5.	Chapra	947.69
6.	Darbhanga	3,036.50
7.	Deoghar	1,721.10
8.	Dhanbad	3,901.42
9.	Dumka	1,351.76
10.	East Champaran	9,153.41
11.	Gaya	2,758.44
12.	Giridih	1,506.30
13.	Godda	725.70
14.	Gopalganj	997.88
15.	Gumla	2,387.32
16.	Hazaribagh	4,265.94
17.	Katihar	4,175.89
18.	Khagaria	4,640.10
19.	Lohardagga	2,170.21
20.	Madhepura	2,741.42
21.	Madhubani	3,743.20
22.	Munger	3,162.78
23.	Muzaffarpur	1,822.10
24.	Nalanda	3,466.82
25.	Nawada	3,184.00
26.	Palamau	2,144.14
27.	Patna	2,176.37
28.	Purnea	3,447.85
29.	Ranchi	3,677.02
30.	Rohtas	924.50
31.	Saharasa	1,057.13
32.	Sahebganj	1,608.80
33.	Samastipur	1,386.13
34.	Singhbhum	4,784.36
35.	Sitamarhi	2,000.00
36.	Siwan	811.66
37.	Vaishali	871.25
38.	West Champaran	4,003.54
	Total area (ha)	95,116.84

even for various domestic purposes. But now a days aquaculture has become their important point of utility.

The ponds and tanks were designed to hold back enormous quantity of monsoon run-off. But unfortunately due to continuous negligence these ponds and tanks are turned from perennial to seasonal due to their physical dereliction. The state of Bihar possesses more than 60,000 ponds and tanks of various sizes ranging from 0.2 to 10 hectares in area, covering a total area of 95,000 hectares (Ahmad and Singh, 1990). These ponds and tanks can be classified into three categories : (*i*) derelict, (*ii*) cultivable and (*iii*) cultivated. The derelict and cultivable ponds/tanks depend on natural stocking and a small quantity of fish is harvested, whereas, the cultivated ponds are being utilized for semi-intensive and intensive aquaculture. The average fish production is only 600 kg/ha. These ponds and tanks are mostly under extensive system, i.e., fish seed is stocked and culture is resorted to without adding fertilizer and feeds. The composite fish culture adopting semi-intensive or intensive system of aquaculture is rarely done for want of seed availability and other inputs. To a large extent, this resource can be developed to cater to the concept of regional self sufficiency in fish requirement.

(ii) 'Mans' (Ox-bow Lakes)

The 'Mans' (Ox-bow Lakes) are water stretches, narrow, depressed, horse-shoe shaped wetlands. These 'Mans' are geologically called Ox-bow lakes and are the old beds of river which at one time formed parts of river meandering in the shape of loops and curves. In course of time, with the shifting and winding course of the flow pattern of the river, the horse-shoe shaped bends are cut-off from the main river and are left in the country side.

Depending upon their age and land contour the two terminal end of the Ox-bow lakes are connected with the parent river through long narrow channel during the monsoon months and become a part of the anastomosis of the riverine system and the riverine spills, they withdraw to isolated existence or completely cut off and remain isolated due to siltation and accumulation of putrified matters in the feeding canals over years.

The lands covered by Ox-bow lakes are not at all suitable for agricultural crops. These water bodies hold enormous quantity of salts water mass. From the point of view of proper land and water utilization, development of Ox-bow lakes deserve urgent attention. But unfortunately these water masses by and large are lying in derelict condition. The development of these water bodies will provide permanent employment to a sizeable population of roughly more than four thousand fishermen families inhabiting along these 'Mans' present. They depend mainly on wild catch of fish and leading miserable life.

The details of the important Ox-bow lakes located in the various districts of Bihar are given in Table 45.2. From the consolidated list, it appears that in the districts of East and West Champaran and Muzaffarpur only there are about 54 Ox-bow lakes covering water spread of 4,229.85 hectares. There are a few more in Vaishali, Samastipur, Darbhanga and Sitamarhi districts. The straight dead river course with a different ecology, called 'Dhars' are present in Kosi basin. The average production of fish from these Ox-bow lakes are estimated to be around 30 kg/ha, mostly consisting of air breathing fishes (Ahmad and Singh, 1991).

(iii) 'Chaurs' (Flood plains)

The plains of Bihar are intersected with vast tracts of problematic land masses classified under the category of 'Chaur' lands and waste lands/ ravines. The 'Chaurs' are floodplains where vast areas remain waterlogged during a major part of the year and in these little agriculture is possible. These vast shallow and low-lying areas which, for want of proper drainage are inundated for greater part of the year and they are neither properly utilized for agricultural operations nor for fish culture. Because of their topography and want of proper drainage, water from the adjoining areas and run-off from long distances stagnate for 4-9 months of the year. Sometimes the 'Chaurs' also receive backwaters during monsoon months from the parent river through the silted-up narrow channels through which these Chaurs are connected at one time or the other with the parent river or with the adjoining Ox-bow lakes in the geological past.

Several Chaurs of such types are available in the district of East and West Champaran, Muzaffarpur, Vaishali, Samastipur, Darbhanga, Madhubani, Purnea, Saharsa, Katihar, Madhepura, Khagaria. Begusaraia and in some parts of Bhagalpur and Sahebganj, covering a water-spread area of over 60,000 hectares (Ahmad and Singh, 1990). Details of some important Chaurs are given in Table 45.3. The Chaurs of Purnea district, however, differ from those of western part of North Bihar in the sense that they are more shallow and retain water in their deepest portions for 4-6 months only. The Chaurs of Gandak region differ from those of Kosi in being deeper and retaining 2-3 feet of water in their deepest portions even throughout the year.

Due to nutrient deficiencies and adverse ecological conditions, the biotic potential of 'Chaur' is very low, as reflected by poor quantum of fish

Table 45.2 : Important 'Mans' (Ox-bow Lakes) of Bihar

	Name of the District	*Name of the Ox-bow lakes and their areas (ha) in parenthesis*	*Total Area (ha)*
1.	East Champaran	Amwa (26), Bakya (160), Basmanpur (40, Bhawanipur (20), Bishambharpur (45), Chakia (20), Chaknaha (400), Chilrion Turkalia (40), Hardia (48), Kararia (100), Majharia-Turkaulia (65), Matwali-Turkaulia (105), Motijheel (100), Paswar (20), Phulwaria (80), Pipra (164), Piprasi (8), Rajpur (80), Rohua Man (20), Rulahi (20), Samnjia (40), Saraiya (80), Sirha Chorwa (200), Sirsa (80), Sonwalla (40), Sugaon (80), Turkauliya (80).	3481.0
2.	West Champaran	Barwalia Tzamali (8), Bhakubar (4), Gahiri (70), Jagarnathpur (40), Karakatti (40), Lal Sariya (230), Mati (40), Narmaida (20), Pasram-Parsa (20), Piprapakri (400), Sajhi-Sathi (40), Sirhachorwa (200), Sombarsa (40).	2,152.0
3.	Sitamarhi	Bamanpura (15), Poaram (30), Ilmasnagar (27).	72.0
4.	Muzaffarpur	Bechaha (30), Bhoosra (45), Brahampura (45), Dholi (8), Gogni (40), Jhapaha (20), Kanti (100), Manika (105.85), Matiha (20), Matipur (110), Murra (15), Rahuwa (30), Rajwara (12), Semera (16).	596.85
5.	Vaishali	Vaishali (40)	40.0
6.	Samastipur	Muktapur (60)	60.0
7.	Darbhanga	Ganga Sagar (19), Dighi (23.20), Harahi (14).	
		Total area (ha)	4,458.05

catch. However, ecologically, these Chaurs have proved to be a favourable habitat for air breathing fishes like Murrels, Magur, Singhi and Kawai and they have their own importance from fisheries point of view.

(iv) Reservoirs (Man-made Lakes)

Impounded water are formed due to the construction of dams, weirs and barrages across the rivers throwing up large expense of waters of higher level upstream of the barrier. These dams have been constructed primarily for the purpose of irrigation, flood control, hydro-electricity generation and water supply.

Table 45.3 : Important 'Chaurs' (Floodplains) of Bihar

	Name of the District	*Name of the Chaurs (Floodplains) and their nearest village/town in parenthesis*		*Area (ha)*
1.	Champaran	Kesaria Chaur	(Motihari)	500
		Chatia Chaur	(Piparapakari)	350
		Manshi Dubey	(Cahur Phulia Khar)	100
2.	Muzaffarpur	Bharthua Chaur	(Bharthua)	125
		Tal Bahaila	(Mehnar)	300
3.	Saharsa	Bhagwa Chaur	(Balua Bazar)	200
		Bora Chaur	(Kharka-Talwa)	500
		Ekpira Dhar	(Kharka-Talwa)	500
		Ekpira Dhar	(Kishan Pur)	200
		Kauda Lauhar	(Kauda Lauhar)	200
		Murdpur Chaur	(Murdapur)	125
		Parbamurli Chaur	(Kumarganj)	100
		Ratnapur Phulkaha	(Kumarganj)	100
4.	Saran	Hardia Chaur	(Akilpur)	10,000
5.	Vaishali	Ahiya	(Rona)	150
		Raghupur Diara	(Hajipur)	2,000
		Fatehpur, Naimallia & Paintia Chaur		11,400
		Total Area (ha)		263,50

The process of obstructing the flow of river water and creating an artificial lake brings a change in the set of ecological parameters. A whole range of physicochemical and biological features are altered. Reservoirs act as silt traps and hence the suspended matter settles down. This results in an increased transparency of water thus results in an increase in primary production. The aquatic ecosystem created after the impoundment of the river offers a great potential for fish production through culture and capture fisheries. Thus culture based fisheries development may be taken up in these reservoirs as a subsequent step towards resource utilization.

Fish and their habitat are considerably affected by river manipulations. Damming resulting in creation of lacustrine condition allows only few fluviatile species of fish to colonize the new environment.

There are big and medium size reservoirs together constituting water spread of approximately 50,000 ha and a few small reservoirs covering a water spread of 15,000 ha extending over a total area of more than 65,000 hectares in Bihar. The list of the most important reservoirs of Bihar are given in Table 45.4. These impounded waters all over the world are developed on the basis of culture based capture fisheries which implies only stocking but no supplementary feeding and fertilization.

II. Lentic (Running) Water Resources

(i) Rivers

The vast network of the various river systems in Bihar mainly dominated by the Ganga river system constitutes the most important capture fisheries resource. Ganga, Kosi, Gandak and Sone together extend a length of 3,200 kilometres. The Ganga river enters Bihar in the middle region, 155 km from Varanasi and covers a total length of 445 kilometres in this State. In this region the Ganga river receives the following tributaries : Ghaghara, Gandak, Burhi Gandak, Sona and Bagmati.

The State of Bihar can be divided into three district zones, the following river basins passes.

1. North Bihar Zone

The river Ghaghara joins the Ganga a few kilometres downstream of Chapra in Bihar. The length of this river is 1,080 kilometres. The Gandak debouches into the plains of Tribeni and flows for another 300 kilometres before it joins the Ganga at Patna. The Burhi Gandak rises from the district of Champaran and covers a length of 320 kilometres. The river joins the Ganga opposite Monghyr. The Bagmati enters Bihar at Muzaffarpur and covers an area of 6,320 sq. km. Kamla enters in Bihar near Darbhanga and joins the Kosi. The Kosi covers an area of 11,000 sq. km. of Bihar and enters near Hanumannagar. After covering 320 kilometres distance below Chatra, it joins Ganga near Kursela.

Table 45.4 : Important Reservoirs (Man-Made Lakes) of Bihar

	Name of the District	*Name of the Reservoirs*	*Area*
1.	Dhanbad	Panchet	7,511.0
		Maithon	11,491.0
2.	Hazaribagh	Tilaiya	6,475.0
		Konar	2,792.0
3.	Giridih	Tenughat	6,000.0
4.	Ranchi	Getalsud	1,400.0
		Hatia	176.0
5.	Palamau	Panghatoa Dam	140.0
		Naudani	276.0
6.	Monghyr	Kharagpur	2.02
		Jhalkund Jalasay	50.0
		Morway Jalasay	40.0
		Nakti Jalasay	143.0
		Nagi Jalasay	469.5
		Amrit Srikhudi	20.0
7.	Bhagalpur	Gadua	1,355.0
		Chandan	1,068.0
		Udaigiri Jalasay	120.0
		Madhyagiri Jalasay	156.0
		Amahara	21.0
8.	Chhotanagpur	Nalkari	992.0
9.	Dumka	Chandan	1,080.0
10.	Santhal Pargana	Canal Dam	10,000.0
		Total area (ha)	51,777.52

	Zones	*Name of the Rivers*
1.	North Bihar	Mahananda, Kosi, Kamla, Bagmati and Adhwara group of rivers, Burhi Gandak, Gandak and Ghaghra.
2.	South Bihar	Karmanasa, Sone, Punpun, Kiul, Badua, Chandan, Gorua, Bhena and Koa.
3.	Chotanagpur and Santhal Pargana	Gumani, Ajoy, Damodar, Subarnarekha, South Koel and Sankh.

2. South Bihar Zone

The Sone basin covers an area of 71,259 kilometres and passes through the Palamau district where it receives the tributary called north Koel. The total length of this river in Bihar is 784 kilometres and joins the Ganga 16 kilometres upstream of Dinapur in Patna. The Punpun rises from Chotanagpur plateau and joins the river Ganga 25 kilometres east of Patna. The length of this river is 200 kilometres and its main tributaries are : Butane, Madar and Morhar. The Kiul with a length of 111 kilometres rises from the Chotanagpur and joins the Ganga near Swaggarha. The main tributaries of this river are : Harhar, Barnar, Azan and Ulan. The river Damodar rises in the south-east area of the Palamau district and cover a length of 541 kilometres. Its main tributary is 'Barakar'. The Rupnarayan rises in the Tilabi hills of Bihar and after covering a distance of 254 kilometres, joins the Hooghly river near Nurpur, downstream of the confluence of the Damodar river.

3. Chotanagpur and Santhal Pargana Zones

The river Subarnarekha rises from Bihar and finally joins the Bay of Bengal. The total length of this river is 395 kilometres. The Karffari river rises from Ranchi region and after covering a distance of 110 kilometres joins the Subarnarekha river near village Chandil. The biggest tributary known as Karkari rises in Mayurbhanj district and joins the river Subarnarekha near Jamshedpur in Bihar. The river Brahmani rises near Nagri village in Ranchi district and is commonly known as South Koel in its upper reaches and covers a length of 260 kilometres in Bihar.

Each of these rivers sustain endemic and some peculiar ichthyofauna. Even the seasonal rivulets like Damodar, Subernarekha and Sankh basin in the plateau region are able to retain a left over assorted population of various fishes in their pools gorges in summer months to procreate and repopulate the stream on the onset of monsoon again. The river system in the floodplains, particularly the Ganga and its tributaries anastomose with vast stretches of canals, abandoned river meanders, lateral pools and depressed lands during rainy season, rendering a mosaic of natural capture fisheries resource.

The estimated fish production from the river is one metric tonne for one mile length of the river stretch (Ahmad *et al.*, 1990). The fisheries of the rivers have to be conserved rather than exploited.

The State of Bihar has a network of irrigation canals in Patnal, Tirhut, Darbhanga and Kosi divisions crisscrossing the country side. These canals retain water only during the irrigation seasons and except for minnows and murrels they do not have much scope for substantial natural fisheries.

References

Ahmad, S.H., Fazal, A.A. and Singh, A.K., (1990) Retrospect and prospects of fisheries in Bihar. Fishing Chimes, February, 1990, pp. 46-49.

Ahmad, S.H. and Singh, A.K., (1990) Scope, prospects, problems and development of food plains (Chaurs) of Bihar for fisheries development. Fishing Chimes, November, 1990, pp. 44-45.

Ahmad, S.H. and Singh, A.K., (1990) Ecological importance of the ponds and tanks of Bihar in relation to fish production. Fishing Chimes, October, 1990, pp. 31-32.

Ahmad, S.H. and Singh, A.K., (1991) Scope, prospects, problems of fishery development of Ox-bow lakes (Mans) of Bihar. Fishing Chimes (In Press).

Bihar State Planning Commission, (1990) Report of the task force on Fishery development in Bihar 2000 A.D. Government of Bihar, Agriculture Division, Bihar State Planning Commission, Patna, Bihar, 28 pp.